DISCARD

COGNITIVE SCIENCE
and
MATHEMATICS EDUCATION

edited by

Alan H. Schoenfeld
University of California

LEA LAWRENCE ERLBAUM ASSOCIATES, PUBLISHERS
1987   Hillsdale, New Jersey   London

Lawrence Erlbaum Associates, Inc., Publishers
365 Broadway
Hillsdale, New Jersey 07642

Library of Congress Cataloging-in-Publication Data

Cognitive science and mathematics education.

Bibliography: p.
Includes index.
1. Mathematics—Study and teaching.   2. Cognition.
I. Schoenfeld, Alan H.
QA11.C54   1987   510'.7   87-5362
ISBN 0-89859-791-9 (hc)
        0-8058-0057-3 (pbk)

Printed in the United States of America
10   9   8   7   6   5   4   3   2   1

*For Anna Louise
with love and hope*

Contents

Contributors

Dr. F. Joe Crosswhite
Mathematics, Northern Arizona University
Flagstaff, Arizona 86011

Dr. James Greeno
Education and Psychology, University of California
Berkeley, CA 94720

Mrs. Anna Henderson
Maury High School
Norfolk, VA 23517

Dr. Jeremy Kilpatrick
Mathematics Education, University of Georgia
Athens, GA 30602

Dr. Steven B. Maurer
Mathematics Swarthmore College
Swarthmore, PA 19081

Dr. Roy Pea
Education, New York University
New York, NY 10003

Dr. Henry Pollak
Bell Communications Research
Morristown, NJ 07960

Dr. Alan H. Schoenfeld
Education and Mathematics, University of California
Berkeley, CA 94720

Dr. Edward A. Silver
Mathematical Sciences, San Diego State University
San Diego, CA 92182

Dr. Ronald Wenger
Mathematical Sciences Teaching and Learning Center
The University of Delaware
Newark, Delaware 19716

Acknowledgments

It is an old and sometimes true saying that "many hands make light work." In the case of this project, the many hands that helped did much more: They made this volume and the conferences that generated it possible in the first place. Support from the Sloan Foundation enabled us to bring together the cognitive scientists, mathematics educators, teachers, and mathematicians whose collaboration resulted in the book. To the Foundation, to all the participants in the conferences, and to the authors go my sincere thanks. Their contributions are evident throughout this book.

The contributions of others are less visible, but equally important. David Spanagel and Margaret Davidson helped to arrange the conferences and to make sure that everything went as smoothly as possible—no easy task during a stormy December in Rochester. Here on the west coast, Jackie Douglass and Jack Smith transcribed the conference tapes. Jane Schoenfeld, my second harshest and best critic, edited the entire manuscript. Jack Smith compiled the indices. (If you haven't done these jobs, you have no idea how tough they are.) The staff at Lawrence Erlbaum Associates has been thoroughly professional, from start to finish, in dealing with the manuscript.

And me? I'm responsible for all the mistakes. Thanks to all the people mentioned above, there are many fewer than there would otherwise be.

Preface

This is an unusual book, produced by an unusual collection of people. Its contributors include professional mathematicians, classroom teachers, mathematics educators, and cognitive scientists — members of groups rarely found in the same room, much less in active collaboration on a substantial project. What brought these people together was a common interest in mathematics instruction and the hope that their combined efforts might help improve it.

Two main observations motivated these efforts. The first is that an emerging discipline called cognitive science is making significant strides in helping to understand "the way the mind works" — and in particular, to understand the nature of thinking and learning processes. Work in cognitive science would, we hoped, be of interest and use to people involved in mathematics education. The second observation is that neither cognitive science nor any of the other disciplines mentioned in the previous paragraph can provide all the ingredients necessary for successful mathematics instruction.

What are some of those ingredients? You need a deep understanding of the subject matter, for without it you may not be teaching "the right stuff." That's one reason for involving mathematicians. You need an equally deep sense of classroom reality to know what's feasible and what's pie in the sky. Classroom teachers live that reality, and mathematics educators study it; both are essential. Finally, you need an understanding of learning and thinking processes. That's where the cognitive scientists come in.

The picture, then, looks something like Fig. P.1. If contributions from all four groups are necessary, then progress in understanding mathematics teaching and learning is most likely to take place at the intersection of the

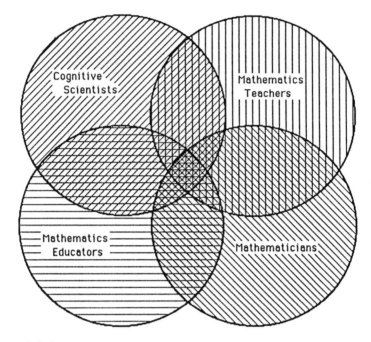

FIG. P.1   Essential Contributors to Progress in Mathematics Instruction

four groups. That's a rather small area. It helps that some people belong to two or three of the groups — as teacher and mathematics educator, for example, or as cognitive scientist and mathematician. However, few people (or even the memberships of standard organizations) have expertise in all four. Our purpose was to build a collaborative effort that allowed us to profit from the *interaction*, rather than the *intersection*, of the four groups (see Fig. P.1).

The need for interaction motivated the two conferences and the extensive follow-up work that produced this book. The first conference was held in June, 1984, in Rochester, New York. Its goal was "foundation building" — establishing lines of communication among members of the different groups, exchanging perspectives, and setting a work agenda for the next 6 months. The agenda was as follows: to explore novel insights about learning, to discuss those findings from the perspectives of the various groups, and to report the distillations of those discussions in clear language that nonspecialists could read and appreciate.

That first conference was no picnic. Members of the different groups represented at the conference don't talk to each other that often; what one group considers central may seem peripheral to another; what one group considers trivial may seem significant to another. People arrived at the conference with different perspectives and, it seemed at times, speaking different languages. Establishing common ground was not easy. Andrea Petitto summed up the experience when she said that we had "exhibited the symptoms of true collaboration: depression, hostility, anxiety, frustration, changes of mind, and glimmers of hope."

That collaboration led to the formation of a number of interdisciplinary working groups (each containing members of all of the different disciplines), which traded papers and ideas over the next 6 months.[1] The task of each group was to try to make sense of a particular topic, and then to report its findings to the group as a whole. Topics included descriptions of theoretical insights and empirical findings from cognitive science, the role of technology in mathematics education, concerns of the classroom teacher, and knowledge about particular domains such as algebra. We convened in Rochester for a second, public conference in December 1984. The main focus of the conference was on the working group reports, which were given by the group leaders. Individual presentations were followed by panel discussions, when members of the working groups reacted to points in the presentations or to questions from the audience. Finally, representatives of the various constituencies — Henry Pollak speaking as a mathematician, Herb Ginsburg as a psychologist, and Joe Crosswhite as teacher and mathematics educator — responded to what they had heard.

This volume reflects the goals and structure of the second conference. Its primary aims are to foster communication among the various disciplines, to reflect the ideas and perspectives expressed by their members, and to convey our distillation of what is new and important in clear, jargon-free language that is accessible to interested readers from any of those areas. Its structure remains true, where possible, to the conference format. Most of the chapters are based on presentations at the conference and are followed by transcripts of the working group discussions that either elaborated on major themes of the presentaions or discussed related issues. The new chapters appear, of course, without additional discusion.

The first three chapters discuss some central themes in cognitive science. Taking up in chapter 1 where this preface leaves off, I provide a broad description of cognitive science — what it is, why it's relevant to mathematics ed-

---

[1]It should be stressed that such exchanges, among all constituencies, were truly bilateral. Though the emphasis here is on the contributions that cognitive science might make to mathematics education, there were many contributions in the other direction. In one notable example, a cognitive researcher mentioned that he had been puzzled by a pattern of students' behavior in a series of laboratory studies. One of the teachers at the conference resolved the dilemma by suggesting what the students' instructional history might have been and how it might have produced the unusual behavior. In this and other examples, teachers' insights and a sense of instruction in context played a central role in our discussions.

ucation, what some of the major implications of research in cognitive science for educators might be. The next two chapters provide detail to flesh out that general introduction. In chapter 2, Ed Silver describes basic work on memory and information processing (limitations of working memory, chunking, etc.). He presents an overview of problem solving theory and research, and describes the instructional implications of the work. Chapter 3 offers Jim Greeno's description of cognitive representations — what they are, how they work and how our understanding of them has implications for mathematics instruction.

Chapters 4 through 6 establish contexts for research and development in mathematics education. In chapter 4, Roy Pea takes a broad look at "cognitive technologies for mathematics education." Arguing that computer-based technologies are merely the latest in a long string of technologies that affect individuals' learning and thinking potential, Pea examines the roles and functions that technology-based instruction can take. In chapter 5, Jeremy Kilpatrick reviews the literature on problem formulation. Though much has been written about problem solving in the past decade, the flip side of the coin — where good problems come from — has not received nearly as much attention. This chapter does some "ground clearing," to set the stage for further work. Chapter 6, "From the teacher's side of the desk," describes classroom reality from the teacher's point of view. This essay by Anna Henderson reminds us of the constraints that the real world imposed on educational practice.

Chapters 7 through 9 summarize particular bodies of work in cognitive science, with direct instructional implications. In chapter 7, Steve Maurer summarizes the literature on "bugs" (students' systematic errors in algorithmic procedures) in arithmetic. In chapter 8, I discuss the literature on metacognition — or, in more down-to-earth language, on "thinking about thinking." In chapter 9, Ron Wenger describes some of the general research on students' difficulties with algebra and how that research led to the development of a computer-based tutorial system for remediation. Finally, reaction papers by Henry Pollak and Joe Crosswhite (chapters 10 and 11) give a mathematician and mathematics educator the last words. Those words are in the spirit of the whole proceedings: they reflect their authors' roots in their disciplines and an effort to provide feedback on our collaborative work from the point of view of the constituent disciplines.

It is important to stress the nature of our collaborations, which were essential to the conference and which resulted in the unique character of this book. The origins of chapter 7 provide a good illustration of the way we came to work together. Like all of our working groups, the group on "bugs" contained members of each of the four disciplines. One of them, Steve Maurer, is a mathematician. He came to the first conference unfamiliar with cognitive science in general and the bugs literature in particular. Another, Kurt vanLehn, is a cognitive scientist who is expert in the area. (It's been the main theme of his research for nearly a decade.) Yet Steve was group leader and author of the report. The reason? The idea was to communicate the results. If Steve could read the work on "bugs" as a mathematician and teacher, discuss it with the group, and communicate its main themes in a way that was satisfactory both to Kurt and the others in the working group, then we had a chance of reaching our audience with the right kind of information. That audience includes:

- teachers interested in the classroom implications of recent research;
- mathematics educators interested in findings from cognitive science relevant to work in mathematics education;
- mathematicians interested in recent developments in research and pedagogy; and
- cognitive scientists interested in the insights of mathematicians, mathematics educators, and teachers into the directions, depth, and ecological validity of recent research in cognitive science.

If you belong to any of these groups, this book should be of interest to you.

1

Cognitive Science and Mathematics Education: An Overview

Alan H. Schoenfeld
Education and Mathematics
The University of California–Berkeley

Twenty-five years ago the phrase "cognitive science" and the field it describes were virtually unknown. Then, at first sporadically and later increasingly through the 1960s and 1970s, an amalgam of researchers from different disciplines, all with common interests in "how the mind works," began to take shape. (For those with an interest in the history of the discipline, Howard Gardner's (1985) *The Mind's New Science* provides a generally accepted outline of the development of the field.) The cognitive science society was formed in the mid-1970s and its journal, *Cognitive Science,* first appeared in 1977. The journal's 1984 self-description, which appears on its inside back cover, provides a good definition of the field and the range of topics that are considered central to it:

> *Cognitive Science* is an interdisciplinary journal. It publishes articles . . .
> on topics such as the representation of knowledge, language processing,
> image processing, question answering, inference, learning and memory,
> problem solving, and planning. . . . [It publishes] theoretical analyses of
> knowledge representation and cognitive processes, experimental studies,
> . . . descriptions of intelligent [computer] programs that exhibit or model
> some human ability, protocol or discourse analysis, . . .

Much or all of the foregoing may seem far from the everyday concerns of mathematics educators. (By "mathematics educator" I mean anyone with a primary interest in the teaching and learning of mathematics. Thus mathematics teachers at all levels and researchers in mathematics education are among those designated by the label.) Indeed, some of the things that cognitive scientists do—for example, spending as many as 100 hours

1

analyzing a single 1-hour videotape of a problem-solving session, and perhaps 2 or 3 years writing computer programs that "simulate" the behavior that appeared in that 1 hour of problem solving—must appear odd to someone looking from outside the discipline. A major goal of this chapter is to demonstrate that such apparently odd behavior can be both sensible and useful. More precisely, my goal is to explicate two main ideas: the idea of a "cognitive process analysis" at a very fine level of detail, and of a "constructivist perspective."

BACKGROUND: SOME ALTERNATIVE PERSPECTIVES AND APPROACHES

A basic assumption underlying work in cognitive science is that mental structures and cognitive processes (loosely speaking, "the things that take place in your head") are extremely rich and complex—but that such structures can be understood, and understanding them will yield significant insights into the ways that thinking and learning take place. Analyses in cognitive science tend to be very detailed. They focus on cognitive processes in an attempt to explain what produces "productive thinking." And because the studies are often carried out in tremendous depth, the number of "subjects" in those studies is often quite small.

The cognitive science perspective is best illustrated by some practical examples, in which the "cognitive approach" can be contrasted with the approaches suggested by more conventional methods. To establish a context for our discussion, we begin with a brief description of some of the learning theories, curricular approaches, and research methods that have had significant impact on American educational practice in this century. Having discussed these, we will turn in the next section to some examples of work in cognitive science that have implications for mathematics instruction.

Associationism

E. L. Thorndike's seminal book *The Psychology of Arithmetic* was published in 1922. Thorndike's learning theory was based on the notion of mental "bonds," or associations between sets of stimuli and the responses to them (e.g. "two plus two" as a stimulus and "four" as a response). According to the theory, bonds become stronger as a result of reinforcement or frequent use, weaker as a result of punishment, or decay as a result of infrequent use. The associationists proposed some general organizational principles for instruction, for example, the principle that bonds that "go together" should be taught together. Translated into pedagogical terms their theoretical approach yielded "drill and practice," a mode of instruction that has had a significant impact on American mathematics instruction. Since the publication of Thorndike's book more than 30 years earlier, drill and practice persisted as a major instructional approach through the 1950s, when I was a student (and used "flash cards" to practice my arithmetic facts). It still exerts a strong influence on the design of many contemporary CAI (computer-assisted instruction) programs. The associationists made some fairly straightforward assumptions about knowledge organization (i.e. about "what's in a person's head" and how it's organized) and had a correspondingly straighforward learning theory. They had little interest in detailed explorations of cognitive structures.

Gestaltism

A different stance was taken by the gestaltists, who believed that mental structures were much more complex than the associationists believed, and that the complexity of such structures needed to be taken into account in teaching and learning. A classic piece of gestaltist exposition is Max Wertheimer's (1959) *Productive Thinking,* originally published in 1945. In it Wertheimer decried rote learning and pointed to the limitations of a drill-and-practice approach. Although such instruction, he conceded, did result in students' "mastering" certain procedures, knowledge acquired in rote fashion was likely to be superficial and thus not likely to be either flexible or useful in a range of situations.

Wertheimer argued his case with a number of pointed examples. A particularly telling one was that many students who were considered to have mastered arithmetic were indeed able to carry out the arithmetic procedures they had studied but had little or no understanding of the meaning of those procedures. Students whom he interviewed would often work problems like

$$\frac{857 + 857 + 857 + 857 + 857}{5}$$

by laboriously adding the five identical terms in the numerator, and then dividing the result by five—all of which is completely superfluous if you understand what the problem calls for. (Wertheimer also quoted a friend's son as saying that he understood arithmetic perfectly well. The child could add, subtract, divide, and multiply with the best of them. The only problem was that the child never knew which method to use.)

The most famous of Wertheimer's examples deals with the "parallelogram problem," the problem of finding the area of a given parallelogram that has base B and height H (Fig. 1.1a). Wertheimer observed a class that had been taught the standard procedure for finding the area, where moving a triangle from one part of the parallelogram to another part creates a rectangle whose area is easy to find (Fig. 1.1b). The students did well and their teacher was proud of their performance. But when Wertheimer asked the students to find the area of parallelograms in nonstandard positions (e.g. Fig. 1.1c) or to find the areas of novel figures to which the same argument applied (e.g. Fig. 1.1d), the students were unable to do so. They (and the teacher) complained that Wertheimer's questions were not fair; the class hadn't studied those kinds of problems.

From the gestaltists' point of view, the questions were fair. The answers to both questions should be apparent if one understands the underlying principles and structures from which the specific arguments the students had memorized could be derived. The gestaltists believed in very rich mental structures and felt that the object of instruction should be to help students develop them. The main difficulty was that the gestaltists had little or no theory of instruction. Although their goal was similar to that of many researchers today (myself included), their theories did not suggest specific instructional methods that could be used to attain it.

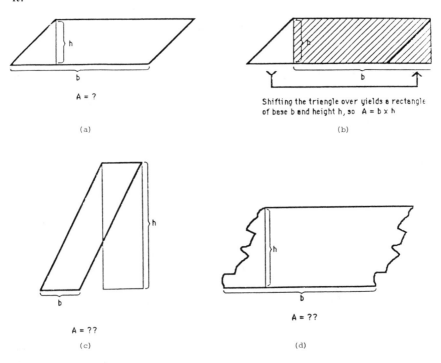

FIG. 1.1. The "parallelogram problem."

Behaviorism

"Radical behaviorists" such as B. F. Skinner (see, e.g., Skinner, 1958) took a stance that was compatible with that of the associationists, but more extreme. In direct opposition to the gestaltists, Skinner held that any emphasis on "mentalism" or attention to "mental structures" was misplaced. He argued that learning performance could be defined solely in terms of observable behaviors ("behavioral objectives") and that learning was best thought of as the result of an individual's interactions with the environment. Thus, behaviorist learning theory focuses on arranging the environment so that optimal interactions take place. Resnick (1983) described the behaviorist approach as follows:

> [Skinner] and his associates showed that "errorless learning" was possible through shaping of behavior by small successive approximations. This led naturally to an interest in a technology of teaching by organizing practice into carefully arranged sequences through which the individual gradually acquires the elements of a new and complex performance without making wrong responses en route. This was translated for school use into "programmed instruction"—a form of instruction characterized by very small steps, heavy prompting, and careful sequencing so that children could be led step by step toward [the] ability to perform the specified behavioral objectives. (pp. 7–8)

Whether in programmed instruction or in other applications, the emphasis on small steps and careful sequencing is central to the behaviorist approach to instruction. This approach, pioneered by Robert Gagné (see, e.g. Gagné & Briggs, 1979), was based on the hypothesis that the right sequence of experiences, repeated with adequate frequency, should generate the right learning. Thus, the bulk of one's attention should be on analysis of subject matter. Gagné focused on constructing careful *task analyses,* which entails decomposing the material to be learned into small building blocks that are mastered individually and later combined into larger units of competency. (I should note that part of the theory included "positive reinforcement" for getting the right answer. Rats and pigeons in Skinner's experimental laboratories were awarded small bits of food when they did well. In an application of the same theory to human learning, my fellow students and I were awarded gold stars when we did well.)

Curricular Trends

For the first half of this century, mathematics curricula in schools were relatively stable. Then, in the late 1950s and early 1960s, our curricula began a series of dramatic swings, each lasting about a decade. The first major shift in curriculum came about in the 1960s in response to a perceived national crisis. The Soviet Union had been the first in space in 1957, with the artificial satellite Sputnik, and the first to achieve manned orbital flight, in 1961. In response to these achievements and to a perceived lack of quality in school curricula, American scientists and mathematicians concentrated intensely on updating and upgrading American science and mathematics instruction. The "new math" and the "alphabet curricula" (BSCS, PSSC, etc.) in the sciences—all with significant mathematical and scientific content—were developed.

A decade later, the new mathematics curricula were generally considered to be failures. The public perception was that school children not only failed to understand the new math, but were no longer able to add, subtract, multiply, or divide. This nationwide reaction engendered the "back to basics" movement. For the next 10 years, teachers once again relied on drill and practice to insure that their students would have basic "foundation" skills in mathematics.

It wasn't long before serious cracks developed in that foundation. It became clear that students were memorizing rote procedures without understanding them, that they were not any better at them than the previous generation of students had been, and that they could not use them in problems that called for even the simplest application. (Evidence documenting these statements may be found in the periodic National Assessments of Educational Progress; see, e.g., Carpenter, Lindquist, Matthews, & Silver, 1983. It may also be found in trends in nationwide SAT mathematics scores, which, despite the emphasis on basic skills in the 1970s, marched steadily downward from 1964 through the early 1980s.) In a major swing of the curricular pendulum, "problem solving" was (re)born in the late 1970s in reaction to the failures of the back to basics movement. In its influential 1977 position paper, the national Council of Supervisors of Mathematics asserted that "learning to solve problems is the principal reason for studying mathematics." Three years later the National Council of Teachers of Mathematics, in its (1980) *Agenda for Action,* made its Recommendation Number 1 that "problem solving must be the focus of school mathematics in the 1980's." The *Agenda* was supplemented by the 1980 NCTM Yearbook, *Problem Solving in School Mathematics* (Krulik, 1980), which was part of a major NCTM effort to support problem solving as the "theme of the 1980's."

We discuss problem solving at some length in the next section. Here we need observe only that most of the curricula just described were based on learning theories whose straightforwardness contrasts sharply with the complex theories of mental structures and learning that are typical of cognitive science. Despite some protests from the gestaltists and some occasional demurrals from the constructivists that instructional practice should consider the richness of mental structures, the primary influence on curricula for the first half of the century was associationism. While the new math was mathematically rich, it was not supported by a comparably rich theory of pedagogy. And the drill-and-practice instruction that predominated during the back-to-basics movement was clearly based on minimalist assumptions about the complexity of mental structures.

Prior Research Methodologies and the "Process *vs.* Product" Distinction

Most of the educational research conducted from 1950 through the late 1970s relied heavily on large-scale statistical methods and the analysis of aggregated data. Such analyses generally proceed under the assumption that patterns of data gathered from large numbers of people provide more reliable information than data gathered from any particular individual. Statistical methodologies have consistently produced reliable and informative findings in disciplines such as agriculture, where the variables in an experiment can be rigorously controlled. Questions such as: Will more frequent watering in small doses, or less frequent watering in larger amounts, produce a larger yield of a particular crop? Or Will fertilizer prove more beneficial when the plants are 3 or 6 weeks old? are ideally suited to them. For this type of question, one creates a set of circumstances that are identical except for the treatment variable(s). Then, when the experiment is performed, differences in outcome—for example, larger *average* yield per plant—can be attributed to differences in treatment. Statistical methods have also provided interesting and useful findings in education, where their use is appropriate—for example, in correlational studies that indicate that students' performance correlates positively with the number of in-service professional development courses taken by their teachers.

The statistical approach, although appropriate and valuable for investigating a wide range of issues, has proved problematic when applied to instructional issues in education. Typical educational research, using "treatment A versus treatment B" comparisons, correlational studies, and factor analyses, was conducted under fairly straightforward assumptions about learning processes—and the assumption that differences in educational "treatments" would show up as statistically significant differences in test results. Unfortunately (for the research, not for the individuals concerned), humans proved to be far more complex than agricultural crops. Through the 1960s and 1970s, large numbers of contradictory or ambiguous studies accumulated in the literature with no resolution in sight. By the mid-1970s, researchers (e.g. Kilpatrick, 1978) were suggesting that statistical methodologies be forsaken for a while, until the research community had a better grasp of the thinking processes it was trying to measure.

The mention of thinking processes in the previous paragraph introduces a fundamental distinction: *process* versus *product*. The focus of process analyses is on the means used to obtain a particular result; of product analyses, on the results obtained. Educational research has traditionally emphasized product, the bottom line being the correct answer or the number of problems a student, or a group of students, can correctly answer. An archetypical example of product-oriented testing is the multiple choice test. In such tests the student's work is not examined to see whether the problems were worked in a reasonable way; the only aspect of the student's work that receives attention is whether or not the correct answer was selected.

Less obvious is the fact that most traditional research studies,[1] aimed at explaining how or why people are good or bad at mathematics, have also been product studies. Consider, for example, typical attempts to explore students' "mathematical abilities." In most factor-analytic studies (see Meyer, 1980, for both a representative study and a literature review), large numbers of students are each given a large variety of tests. The results are then analyzed statistically, to reveal patterns of regularity in test scores. In Meyer's study, for example, 179 fourth-grade children each took 22 tests ranging from arithmetic tests to "figure matrices" to a problem-solving test. Data analyses revealed six different factors (one verbal, two inductive, one numerical, one perceptual speed, and one genral mathematics factor) in mathematical ability. Hypothetically these factors revealed information about individual performance: An individual student's scores on each factor would reveal underlying strengths or weaknesses and thus predict future success or difficulty in particular endeavors. Similar methodologies have been used to explore the roles of visual spatial abilities and sex differences in problem solving. In general, "having an ability" has been defined as scoring well on a test for that ability. Thus, statements like "verbal ability is an important aspect of problem solving performance" really meant "scores on tests of verbal ability correlated highly with scores on scores on a problem-solving test." For the most part, classic educational research that claimed to explore "problem-solving performance" did not examine what people were doing—or trying to do—when they worked problems. In contrast, examining what people do in their problem-solving attempts is central to much cognitive science work on problem solving. Cognitive scientists looking at

---

[1]By "traditional research studies" I mean the vast majority of studies appearing in educational journals through the late 1970s. See, for example, typical articles in the *Journal for Research in Mathematics Education* over the past decade. Things have been changing, with a clear swing toward process-oriented studies in recent years.

issues of learning and problem solving ask a very different series of questions than do classic educational researchers, and they use very different methods to answer them. Such methods are the focus of the next section.

ILLUSTRATIONS OF WORK IN COGNITIVE SCIENCE

We begin this section with a broad, general description of cognitive analyses. This description continues the discussion of the previous section, indicating how a cognitive scientist might approach issues of specific "mathematical abilities." We then focus on two specific examples, drawn from Bundy's (1975) work on equation solving and my (1985) work on problem solving.

Aspects of the Cognitive Science Approach

There is no single methodology, or set of methodologies, characteristic of work in cognitive science. There are, however, some features typical of cognitive analyses. To illustrate some of these features, we shall indicate the way that a cognitive researcher (in contrast to the factor-analysis researchers described earlier) might explore the role of verbal ability in mathematical performance. This researcher might choose some representative verbal ability tasks, probably including tasks that appear on standard verbal ability tests. A few students would be asked to work the problems out loud, and videotapes would be made as the students did so. These videotapes and the written work produced by the students would be analyzed in detail, with the intention of identifying consistent behavior patterns. Other kinds of tests might be used to probe the students' knowledge and the way it was organized.

The cognitive scientist might then try to construct a "process model" of the students' understanding. This means specifying the particular knowledge accessible to the student, the thinking strategies the student appears to have available, and the nature of the interactions between the two. Different kinds of models might be constructed, possibly simulation models (attempts to specify the knowledge and behavior of particular students) or idealized models (attempts to specify the knowledge and behavior that produce competent performance.) Then the cognitive scientist might construct a computer program designed to behave in accord with the model. Theoretically, the idealized program would solve the "verbal ability" problems correctly. Theoretically, the simulation program not only would produce the same answers as the student, but would work the problems in the same way the student worked them.

At this point the cognitive scientist may seem to have gone rather far afield, especially since writing the type of computer programs described in the previous paragraph may take thousands of hours. What do these programs have to do with the student's behavior?

A great deal. The idealized model reflects a hypothesis of the following type: Using certain kinds of knowledge (which are specified in detail), and certain kinds of thinking strategies (which are comparably well specified), one can solve certain kinds of problems. The computer program puts that model to the test by performing the processes that hypothetically yield success. The program has the knowledge and the strategies. If it does solve the problems, the result is a "sufficiency proof": the ideas do work. If the program fails, the scientist has still obtained valuable information: The failure demonstrates that the theoretical model was inadequate. The scientist may have thought that the knowledge and thought patterns required to solve problems in that particular domain had been adequately characterized, but the fact that the program did not succeed (or worked differently than expected) indicates that they were not fully understood. With the empirical evidence of the model's weakness provided by the computer program, the scientist can go back and try again—but with greater understanding.

Research using simulation models works in much the same way. A simulation program tests the hypothesis that a particular kind of knowledge and knowledge organization, and particular thinking strategies, produced the individual's behavior. If the hypothesis (and the programming) are correct, the program's behavior will parallel the student's. If the program's behavior does not match the student's, there is clear proof that the model is inadequate. Hence the computer programs offer empirical tests of theories that attempt to characterize the nature of thinking— theories that are often so complicated that the only way to test them for certain is to try them out in this way.

For the historical record (and to help establish a context for the discussion of my work to come), it should be noted that much work along the lines sketched earlier is based on work in Artificial Intelligence (AI), the field that tries to construct "machines who think" (McCorduck, 1979). A classic work in AI is Newell and Simon's (1972) *Human Problem Solving,* in which the genesis of a computer program called General Problem Solver (GPS) is described. GPS was developed to solve problems in chess, symbolic logic, and "cryptarithmetic" (a puzzle domain similar to cryptograms, but with letters standing for numbers instead of letters). GPS played a decent game of chess, solved cryptarithmetic problems fairly well, and managed to prove almost all of the first 50 theorems in Russell and Whitehead's *Principia Mathematica* (1925–

1927)—all in all, rather convincing evidence that the problem-solving strategies in GPS were pretty solid.

Where did those strategies come from? They came from detailed observations of people solving problems. Newell and Simon recorded the attempts of many people to solve problems in chess, symbolic logic, and cryptarithmetic. They then examined those attempts at an extraordinarily fine level of detail, looking for uniformities in the problem solvers' behavior. To borrow the relevant term from chess, the possibilities for each "move" were examined systematically. Questions like the following were examined: Which moves were explored, and how far were they pursued? What criteria determined which moves to pursue, and to what depth? If the position being examined resembled another position, was a similar move chosen in that position? If so, why (i.e. what was it about the similarity that called for the same move); if not, why not? Similar observations were made for subjects' strategic moves in cryptarithmetic and symbolic logic, and the three domains were compared. Were there similarities across domains? Were there regularities in behavior? Could those regularities could be described precisely and written as computer programs? And would the program succeed in solving problems in the three problem domains? The evidence, described earlier, speaks for itself. Newell and Simon's anslyses produced a set of *prescriptive* procedures— problem-solving methods described in such detail that a machine following their instructions could obtain impressive results. The notion of prescription, rather than description, is central to our discussion of mathematical problem solving. So is the notion of "grain size," or the level of detail of the analyses that one performs.

An analogy may be useful to clarify the concept of grain size. Consider the case of biologists interested in analyzing the physical traits that are passed by one generation of organisms to the next. Following in the Mendelian tradition, one can perform analyses at the level of genotype: for example, when a black (BB) Andalusian fowl is crossed with a splashed white (bb) Andalusian fowl, all the offspring are hybrid (Bb), popularly called "blue" Andalusian fowl. When two blue Andalusian fowl are crossed, one can expect 25% of the offspring to be black, 50% to be blue, and 25% to be white. Such analyses, usually accompanied by Punnett Squares or by diagrams indicating gamete combinations, are true process models with a long history of successful use. Another kind of process model, describing the same phenomena, operates at the level of molecular genetics. Open a recent issue of *Science* and there is a good chance that if the genetic properties of a black andalusian fowl (or any other organism) are discussed, the characterization of the organism will consist of a page-long series of As, Cs, Ts and Gs, representing its DNA macromolecular base-pair structure. This will be followed by a detailed discussion of that structure, possibly by a discussion of what happens when a small segment of the molecular chain is replaced by a similar, but slightly altered segment. The molecular analyses are similar to intent to the Mendelian analyses, but at a much finer grain size: the fundamental units employed in the analyses are much smaller. The issue is not whether one analysis is right and one wrong, for the analyses are compatible. The issue is whether analysis at one level or the other is more informative (produces better and more useful predictions).

Similarly, analyses of thought processes can be performed at various levels. Work in cognitive science tends to be very detailed in contrast to the work described in the first section of this chapter. Consider base 10 subtraction, for example. One level of analysis is to consider the subprocedures required to carry out the algorithm. Does the student know the relevant number facts and the procedure for borrowing? An analysis of a student's work at this level, typical of behaviorist task analyses in the 1960s, could isolate some problematic aspects of the procedure. A more detailed analysis describes the sequence of observations and operations necessary to carry out the subprocedures, such as: noting the two digits in a column; comparing them to see which is larger; using number facts for the subtraction if the top number is larger and writing down the resulting digit, or initiating the borrow procedure if the smaller is larger; using the standard borrow procedure if the top digit in the column to the left is not zero; using the special borrow procedure if the top digit in the column to the left is zero; and so on. One such analysis appears in Brown and Burton (1978). Figure 1.2 illustrates the level of detail in Brown and Burton's analysis.

More detailed process analyses look at all the places where something can go wrong when a student tries to implement the procedures and trace the possible consequences of such mishaps. Even more detailed analyses may include representations of the ways that number facts are stored or recalled,[2] and the procedures for recalling them. Yet, as chapter 7 indicates, such research has clear and practical implications for mathematics instruction. Another example with such implications for instruction is found in chapter 3. There Jim Greeno describes the "semantic

---

[2]A *representation* is a symbolic description of objects and relations among them. For example, the string of As, Cs, Ts and Gs that are used to characterize DNA macromolecular base-pair structure are a representation, with each letter standing for a particular type of molecule and the sequence of letters in the string standing for a particular organization of those molecules. Note that a representation does not (necessarily) capture all aspects of the objects being described: the string of letters says nothing, for example, about the helical structure of the DNA molecule. Thus a particular representation may be useful for a particular line of inquiry, but not for another.

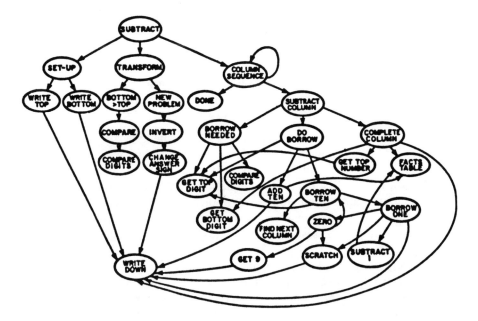

FIG. 1.2. A procedural network for subtraction. From Brown and Burton (1978). Reproduced with permission.

network" that represents solutions to simple word problems such as "Jay had nine books; then he lost four books. Now how many books does he have?" The diagrams representing such problems become fairly complex, and these diagrams are only the tip of the iceberg: The computer programs that simulate students' work on the problems are far more complicated. At first glance, these programs may appear to be as far removed from classroom discussions of word problems as typical research in molecular genetics appears to be removed from the concerns of animal husbandry. That first glance is wrong in both cases; see chapter 3 for details about word problems.

A Specific Example: Bundy's Work on Equation Solving

As an illustration of the cognitive science approach we consider a particular problem domain, solving complex algebraic equations. Most of our discussion is based on Bundy's (1975) work. Ron Wenger picks up this discussion and extends it—also with direct classroom applications—in chapter 9.

If one takes the "subject matter" point of view, the task of algebraic equation solving can be decomposed into a collection of typical problem subdomains and their associated solution methods. There are, for example, standard procedures for:

linear equations of the form $ax + b = cx + d$;
quadratic equations of the form $ax^2 + bx + c = 0$;
complex polynomial equations;
rational algebraic functions;
equations involving radicals (see example below);
logarithmic equations of various forms (see example below);
trigonometric equations of various forms, etc.

From the behaviorist "task analysis" perspective (e.g. Gagné & Briggs, 1979), this catalogue (or a more detailed version of it) suggests the appropriate instructional approach. Students could be taught solution methods for the individual types of equations and given practice until they have demonstrable competence. Hypothetically, as students build up competence in each domain, the sum total of those domain-specific competencies would result in a general mastery of equation-solving procedures.

Indeed, most students learn the techniques in individual domains in just that way. However, recent cognitive research suggests that the resulting general competencies (when they occur) are quite different from the sum-of-the-parts model that the behaviorists propose. Bundy (1975) made a detailed study of competent mathematicians solving algebraic problems like

$$\log_e(x + 1) + \log_e(x - 1) = 3,$$

$$\sqrt{5x - 25} - \sqrt{x - 1} = 2, \text{ and}$$

$$\arcsin(3x) + \arccos(2x) = \arcsin(x)$$

The study was motivated by the observation that mathematicians' work on such equations is far more efficient than the sum-of-the-parts model would suggest. Indeed (since the techniques for the various domains listed in the foregoing are essentially algorithmic), the various domain-specific pieces had been implemented as computer programs. Larger, sum-of-the-parts programs were also built. However, despite their reliable, quick, and error-free performance on typical problems from particular subdomains (more than we can expect of students!), programs built along those lines came nowhere near matching competent human performance in general. They could not match the directness of human solutions—rapidly finding the "right" solution methods for complicated problems—or the range of problems solved. Bundy (1975) set out to explore a competent mathematician's equation-solving skills. "In order to try to match the human mathematician's performance we analyze his solutions. This analysis reveals a system of high level strategies and goals, which can be used to guide the search for a solution . . ."(p.5).

The focus of the study, then, was on how a *particular individual* solved problems. That person's solutions were examined, line by line, for the purpose of understanding what the mathematician did, and why. The nature of the explanations extended beyond subject matter techniques to an elaboration of the mathematician's thought processes. They focused on *strategies* used by the individual—strategies in the service of certain *goals*, which helped him make progress towards a solution.

Two of the higher order strategies attributed by Bundy to the mathematician[3], and used for problems as different as the three problems illustrated earlier, can be paraphrased as follows: (a) collect like terms to simplify expressions, and (b) try to replace "nasty" terms with nicer ones. Those strategies can be applied to the first equation,

$$\log_e(x+1) + \log_e(x-1) = 3,$$

as follows. Collecting terms first yields

$$\log_e(x+1)(x-1) = 3,$$

and then

$$\log_e(x^2-1) = 3.$$

Since exponentials are "nicer" than logarithms, the conversion to

$$x^2 - 1 = e^3$$

is indicated. Adding 1 to both sides of the equation (i.e. isolating x^2) and taking the square root of both sides produces the solution.

For the preceding problem the strategies seem "natural," and their use is compatible with the domain-specific approach that would be taken by the behaviorists. Each step yields an equation that is obviously simpler than the preceding one; thus there are clear signs of "progress" at each step. This is not the case in the second and third problems, where the difference between the behaviorist and cognitivist approach becomes clear. In the second problem,

$$\sqrt{(5x-25)} - \sqrt{(x-1)} = 2,$$

radicals are "nasty" whereas polynomials are nice(r). This suggests squaring the equation (perhaps after shifting one radical to the other side of the equals sign). Squaring the original yields

$$(5x-25) - 2\sqrt{(5x-25)(x-1)} + (x-1) = 4,$$

---

[3]This is indeed an attribution, in inference made on the basis of the mathematician's systematic behavior. Bundy (1975) does not assert that the mathematician is consciously and systematically applying those strategies.

which is more nasty looking than the original. Further simplications, which require algebraic regrouping before squaring for a second time, are needed. Only then is the resulting equation "nice," that is, a simple polynomial equation that can be easily solved. Hence planning and a willingness to tolerate temporary chaos in the service of long term goals— the *eventual* simplification of the equation—constitute part of the mathematician's competence.

The third problem,

$$\arcsin(3x) + \arccos(2x) = \arcsin(x),$$

provides a more dramatic demonstration of this point. There are three nasty terms: arcsin(3x), arccos(2x), and arcsin(x). Inversion may simplify matters, but since there are three candidates, which function should be inverted? One can reason that inasmuch as sin(3x) and cos(2x) can be expressed easily in terms of sin(x), arcsin(x) is the most likely candidate. Performing the relevant inversion yields the equation:

$$\sin(\arcsin(3x) + \arccos(2x)) = x,$$

which can be *expanded* by the identity for sin(A + B) into

$$[\sin(\arcsin(3x))][\cos(\arccos(2x))] + [\cos(\arcsin(3x))][\sin(\arccos(2x))] = x.$$

The first two terms in brackets are now easy to evaluate. Although the next two may look horrendous, there is a standard technique for determining their algebraic values. The equation can thus be converted into

$$(3x)(2x) + \sqrt{(1-9x^2)(1-4x^2)} = x.$$

The rest, as they say, is algebra.

Some comments are appropriate before we move on. We should note that the strategies used by the mathematician cut across problem types: "getting rid of nasty functions" applies to equations with radicals, with trigonometric functions, with logarithms, and so on. This type of strategy is not domain specific, as are the techniques for solving problems in each of those areas. In a serious sense, such strategies transcend the domain-specific techniques. With them one can solve problems that have not been encountered before and for which one does not have prepackaged solutions. Moreover, many of these strategies may not have been explicitly taught. Thus they represent an integration and restructuring of equation-solving knowledge: what the mathematician knows about solving equations is not only more than, but different from, a collection of the techniques mastered in individual domains. To recapitulate some main themes of this discussion, we note the following. First, Bundy's (1975) analysis focused on process and strategy, on what the mathematician did while working the problems. Second, Bundy worked with a small number of subjects (in this case one) and made extremely detailed analyses of the person's work. This type of approach is predicated on the assumption that mental structures are quite complex and need to be modeled in some detail. Third, his approach was prescriptive: The intention was to specify the strategies in precise enough detail so that you could try them and tell (for certain, in the case of computer implementation) if the strategies really worked.

A Second Example: Some of My Work in Problem Solving

As I noted in the section on mathematics curricula, two major swings of the mathematical curricular pendulum in the 1960s and 1970s resulted in "problem solving" becoming the theme of the 1980s for school mathematics. Whereas the theme may have been pegged to the 1980s, the roots, the rationale, and the methodologies of the problem-solving movement were solidly based in the 1940s—in 1945, to be precise. That was the year that George Pólya (1945/1957) produced his extraordinary little book *How to Solve It.* The book was a *tour de force,* a charming exposition of the problem-solving introspections of one of the century's foremost mathematicians. (If you don't own a copy, you should.) Like Descartes 300 years before him, Pólya examined his own thoughts to find useful patterns of problem-solving behavior. The result was a general description of problem-solving processes, a four-phase model of problem-solving— understanding the problem, devising a plan, carrying out the plan, and looking back—the details of which included a range of problem-solving *heuristics,* or rules of thumb, for making progress on difficult problems. The book and Pólya's subsequent elaborations of the heuristic theme (e.g. Pólya, 1981) are brilliant pieces of insight and exposition. Mathematicians generally agree that Pólya's descriptions of problem-solving strategies are accurate; the mathematics education community has adopted Pólya's approach as *the* approach to problem solving; and virtually all of the research on mathematical problem solving in recent years has been based on Pólya's work.

Despite the brilliance of Pólya's insights, the methods suggested in his books did not work very well; when teachers and mathematics educators tried to teach problem solving using Pólya-like methods, the results were equivocal at best. Study after study in the math-ed literature produced "promising" results, where teacher and students alike were happy with the instruction, but where there was little evidence, if any at all, of improved problem-solving performance. The reason is as follows. Pólya's characterizations of problem-solving strategies were in essence accurate summary descriptions. People who already knew the strategies (for example, the mathematicians who read his books) could and did recognize them in Pólya's descriptions. But there is a huge difference between *description,* which merely characterizes a procedure in sufficient detail for it to be recognized, and *prescription,* which characterizes a procedure in precise enough detail so that the characterization serves as a guide for implementing the strategy. Accurate though they were, the descriptions of problem solving strategies in *How to Solve It* did not provide enough information to enable the student to use them. For the strategies to be usable, it was necessary to create prescriptive versions of the strategies, at the right (i.e. implementable) level of detail.

My approach to that task was inspired by classic work in artificial intelligence, some of whose methods were described in the discussion of Newell and Simon's (1972) *Human Problem Solving*[4] and Bundy's (1975) work. My intention was to pose the question of problem-solving heuristics from a cognitive science perspective: What level of detail is needed to describe problem-solving strategies so that students can actually use them? Methodologies for dealing with this question were suggested by those used in artificial intelligence. One could make detailed observations of individuals solving problems, seek regularities in their problem-solving behavior, and try to characterize those regularities with enough precision and in enough detail that students could take those characterizations as guidelines for problem solving.

That suggestion is, of course, a gross oversimplification. Even if the strategies could be described with some degree of accuracy, one certainly couldn't "program" students like computers. But the metaphor had some force. I made detailed studies of good problem solvers and looked for regularities in their behavior. It turned out that there were regularities—and that they were at a much "finer" level than Pólya had described. To pick a simple strategy as an example, Pólya was right that mathematicians examine special cases as a way of making sense of problems and gaining intuition about them.[5] That strategy, described in general, reads as follows: To understand an unfamiliar problem better, you may wish to exemplify the problem by considering various special cases. This may suggest the direction of, or perhaps the plausiblity of, a solution.

These 31 words provide a convenient summary statement of the approach. Unfortunately, there is hardly enough information contained in them, even when exemplified on a reasonable collection of problems, to enable students to use the approach. The reason is that there are, in

---

[4]For a very readable introductory discussion of such work and its implications for mathematics educators, see Rissland (1985).

[5]For an extended discussion of this topic, see chapter 3 of my (1985) *Mathematical Problem Solving.*

reality, a fair number of rather distinct special-cases substrategies. Five of those strategies are listed below:

1. If there is an integer parameter n in a problem statement, consider the values $n = 1,2,3,4, \ldots$ You may see a pattern that suggests an answer, and the calculations themselves may suggest the mechanism for an inductive proof that the answer is correct.
2. If a problem asks about the roots of polynomial equations in general, look at easily factorable polynomials. Similarly for other functions: when a problem is about zeros of arbitrary functions, choose special cases whose zeros are easy to trace.
3. In problems where terms are defined recursively, try letting one of the initial terms be 0 or 1; the resulting calculations may be more straightforward, allowing you to see what's really going on.
4. A range of "without loss of generality" arguments.
5. Examine conjectures on "nice" polygons instead of arbitrary ones: on rectangles or squares instead of arbitrary quadrilaterals; on isosceles or right triangles, and so on.

Similar statements can be made regarding most of the strategies described in *How to Solve it* (Pólya, 1945/1957). Detailed analyses revealed that any particular "strategy" described by Pólya was in reality not one, but five or ten different strategies. Process analyses revealed how the methods "worked" and allowed me to create *prescriptive* versions of the strategies. ("Under the following conditions, you should try the following things, in the following ways.") In that way, the methods of cognitive science suggested what I might try to take into the problem-solving classroom and how I might try to teach some aspects of problem solving.

The story of what I did and how it worked is too long to tell here, but one result is worth mentioning. The final examinations for my problem-solving courses had three parts. Part 1 contained problems similar to problems we had worked in the course and on which the students had been "trained" to use problem-solving strategies. Part 2 contained problems that could be solved by those strategies, but which did not resemble in any obvious way the problems we had worked in the course. Part 3 consisted of problems that had stumped me. I had looked through contest problem books, and as soon as I found a problem that baffled me, I put it on the final exam. The students did remarkably well on parts 1 and 2, as I had hoped. But they also did well on Part 3: A number of students solved problems I had not solved when I handed out the exam.

Of course, the story is hardly that simple: see my (1985) *Mathematical*

Problem Solving for details. But one of its "morals," parallel to one in Bundy's (1975) work, bears repeating. By making detailed analyses of the ways that people solved problems, I was able to characterize some useful problem-solving strategies, and the strategies were defined precisely enough so that students were able to use them. Asking the question, What does it take to make this strategy work? was useful in both domains. The question, and the methodologies for answering it, typify some of the work currently being done in cognitive science.

The Constructivist Point of View

Constructivism is not a part of cognitive science *per se;* it is, rather, a perspective that has come to play an increasing role in cognitive inquiries and that promises to reshape many of our ideas about learning and teaching. My purpose in this section is to outline what might be called the constructivist perspective and to point out some of its implications for instruction.

Since Piaget is the most famous of the constructivists, we establish the context for our discussion with two brief examples take from Piagetian studies. In the first, Piaget (1954) reports his interview with a 9½-month-old girl.

> Jacqueline is seated and I place on her lap a rubber eraser which she has just held in her hand. Just as she is about to grasp it again I put my hand between her eyes and the eraser; she immediately gives up, as though the object no longer existed.
>
> The experiment is repeated ten times. Every time that Jacqueline is touching the object with her finger at the moment when I cut off her view of it she continues her search to the point of complete success . . . On the other hand, if no tactic contact has been established before the child ceases to see the eraser, Jacqueline withdraws her hand. (p. 22)

Piaget stresses that it is not a shift of interest (i.e. a focus on his hand instead of the eraser) that causes the child's forgetfulness. Rather, "it is simply because the image of my hand abolishes that of the object beneath it, unless . . . her fingers have already grazed the object or perhaps also unless her hand is already in action under mine and ready to grasp" (p. 23). This episode illustrates that children do not have the same sense of object permanence—the understanding that objects continue to exist when they leave our field of vision—that adults have. Piaget's research indicates that "out if sight, out of mind" may indeed be an accurate description of cognitive reality for very young children; it takes about 18 months for infants to develop a sense of object permanence similar to that of adults.

The second, and more famous, Piagetian example deals with "conservation of volume." Adults know that the volume of a liquid remains the same when the liquid is poured from one container into another: An 8-ounce can of soda will yield 8 ounces of soda, whether the soda is poured into a tall, thin glass or a short, wide one. As we know, young children do not see things the same way. When equal amounts of soda are poured from identical glasses into glasses of different shapes (see. Fig. 1.3), children will say that there is more soda in the glass that has a higher level of liquid; given the choice of drinking from either glass, they will consist-

When identical amounts of soda are poured from identical glasses ...

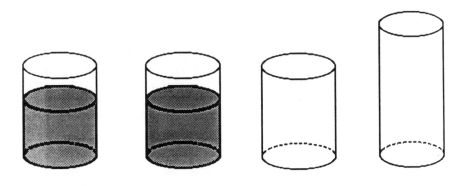

into differently shaped glasses ...

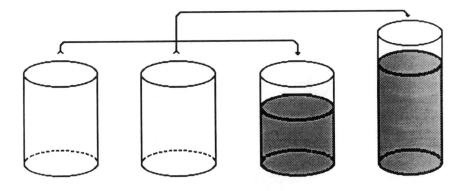

younger children will say there is more soda
in the glass with the higher level of liquid.

FIG. 1.3.

ently choose the one with the higher level of liquid. (See Ginsburg & Opper, 1969, for an introductory discussion of this and related experiments.)

For the purposes of this discussion, the significance of the Piagetian examples is that they highlight a conflict between two different versions of reality. On one hand, there is the adult perception of "real-world" phenomena. We take object permanence for granted: When a ball rolls under a couch we expect to find it there, unless it has rolled out the other side. We also take conservation of volume to be an obvious fact of nature. Simply put, object permanence and conservation of volume are objective facts from the adult's point of view; they describe the world "as it is." On the other hand, those "objective realities" are seen quite differently by children. The very young child lives in a world where some objects exist and then cease to exist; a doll covered by a blanket simply disappears, and its disappearance is not a "problem." Somewhat older children live in a world where volume is not conserved; transformations like pouring not only change the shape of a liquid but change its volume as well. Children do not see the world as we see it. Although they may observe the same phenomena we observe, they interpret them differently. Children's interpretive frameworks, their own constructions of reality, determine what they see.

In fact, the same is true for adults. According to the constructivist perspective, we all build our own interpretive frameworks for making sense of the world, and we then see the world in the light of these frameworks. What we see may or may not correspond to "objective" reality. To illustrate this point, we may consider any optical illusion, for example the one given in Fig. 1.4. Even after we measure the two horizontal lines to verify that they are the same length, it is difficult to see

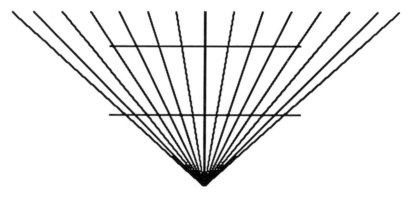

FIG. 1.4.   An optical illusion indicating the power of interpretive frameworks.

them that way. What we "see" in the illusion is an *interpretation*, an interpretation that in this case contradicts a demonstrated fact.

For a more serious illustration we turn to a domain closer to mathematics. There is now an extensive body of research in a field called "naive physics." McCloskey (1983), for example, studied the kind of situation illustrsted in Fig. 1.5. Figure 1.5 shows a thin curved metal tube. You are looking down on the tube, which is lying flat on a table. A metal ball is inserted into the tube at the end, indicated by the arrow, and is shot out of the other end at high speed. You are asked to ignore air resistance and spin and then draw the path that the ball follows after emerging from the tube.

The correct answer is illustrated in Fig. 1.6a. The answer follows from Newton's first law, which states that in the absence of net applied forces, an object in motion will move in a straight line. A frequently obtained incorrect answer is illustrated in Fig. 1.6b. In a series of studies (Caramazza, McCloskey, & Green, 1981; McCloskey, 1983; McCloskey, Caramazza, & Green, 1980), this answer was given by half the college students who had not studied physics in high school, about a third of those who had, and about a seventh of those who had studied physics in college.

In related work, diSessa (1983, 1985) documents a number of consistent misinterpretations of physical phenomena and provides a theory of their origins. One false intuition, "force-as-mover," is described as follows: A "force causes motion in the direction of the force, ignoring the effect of previous motion" (diSessa, 1983, p. 30). This intuition is illustrated in Fig. 1.7. A person who employs force-as-mover would not make the correct prediction illustrated in Fig. 1.7a, but would instead predict the motion described schematically in Fig. 1.7b. Other false intuitions include the beliefs that forces die out on their own and that a constantly applied force is required in order to keep an object moving at constant speed.

Each of these misconceptions is plausible, and in a limited sense each is correct. We encounter force-as-mover when we push an object that is at rest. In that case, the object does indeed go in the direction of our push. Moreover, force-as-mover seems to work in other circumstances: We seem to push a lawn mower in the direction we want it to go, even when we push it around corners. From these experiences and similar ones, it

FIG. 1.5.

a. The Correct Answer   b. A Typical Incorrect Answer

FIG. 1.6.

seems reasonable to abstract force-as-mover as an explanatory framework and to employ that framework to explain like circumstances.
(Neither the abstraction process nor the means of accessing the framework is presumed to be conscious.) Similarly, a law of "dying forces"
seems plausible. All the objects we have pushed have come to rest; the
noise of a bell fades with time; in fact, any process we set in motion comes
to a halt. (This last generalization is one way to interpret the statement
that "it is impossible to construct a perpetual motion machine.") Finally,
we observe that it appears necessary to exert constant effort to keep a
lawnmower moving, that it is necessary to keep one's foot on the gas
pedal to keep a car moving at constant speed, and that running requires
consistent effort. These experiences may (again, not necessarily consciously) result in one's forming the impression that constant force is
required to produce motion at constant speed. In short, the misconceptions identified in the naive physics literature are quite natural. They are
obvious, but incorrect, generalizations of one's experience. They are
also, as the literature indicates, quite robust. Many students enter physics
courses "seeing" phenomena in ways that violate the principles of
physics that they will study, and many of them leave with those misconceptions intact.

To recapitulate, the constructivist point of view is the following;
"Human beings are theory builders; from the beginning we construct
explanatory structures that help us find the deeper reality underlying
surface chaos" (Carey, 1985, p. 194)." From infancy to the grave, people
construct their own views of reality, views that may not always coincide
with what some would call objective reality. Because our experiences
with the world are more or less homogeneous, we tend ultimately to
develop the same perspective, as in the case of object permanence and
conservation of volume. But everything we see is an *interpretation*. As
the optical illusions and physics misconceptions discussed earlier illustrate, these interpretations can be incorrect. What we "see" may be not
be what is actually "out there," but rather the results of consistent
misinterpretations. Moreover, the explanatory frameworks that produce
those mininterpretations may be very resistant to change.

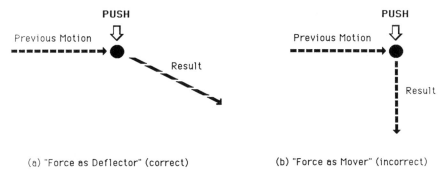

(a) "Force as Deflector" (correct)   (b) "Force as Mover" (incorrect)

FIG. 1.7.

Though the examples in this section were taken from topics outside mathematics, the implications of the constructivist perspective for mathematics instruction are all too clear. A large part of our teaching practice is based on incorrect assumptions, in particular the assumptions that characterize the *absorption model* of instruction: "The mathematics [we teach] is assumed to be a fixed body of knowledge, and it is taught under the assumption that learners absorb what has been covered" (Romberg & Carpenter, 1986, p. 26). In other words, most teaching assumes that the student is a blank slate, a *tabula rasa*. We show the students a procedure (how to subtract, how to factor, how to solve simultaneous linera equations, how to differentiate using the chain rule) and hope the message gets across. Since the message doesn't usually take on the first try, we show them again. If we are inventive, we show them a slightly different version the second time. By the time we're done, we may have shown them the same thing five or six different ways—all in the hope that the message will eventually sink in.

This approach makes good sense; it tries to reach the student in many different ways. But it also makes a fundamental error in assuming that the student who hasn't learned a topic yet simply "hasn't learned it yet." It may be the case, as with misconceptions in physics, that the student has indeed learned something—a consistent interpretation of the subject matter that just happens to be wrong. In that case our explanations of the right procedure, no matter how clever, may fall on deaf ears. The student may see what we do and interpret it differently, just as the preconservation child sees a very different event that we do when a liquid is poured from one container into another.

There are numerous comparable examples in mathematics, as chapter 7 indicates. There Steve Maurer reviews the research on arithmetic "bugs," students' consistent misinterpretations of simple procedures such as addition and subtraction. The research indicates that at least a third, and perhaps as many as half, of students' mistakes on addition and subtraction problems are not simply random. They are systematic errors, so systematic, in fact, that it is often possible to predict an individual student's incorrect answers before the student even works the problems! Let me not steal Maurer's thunder: the details are in chapter 7. I shall, however, point to a few other examples of where students' naive expectations from outside mathematics, or their incorrect abstraction of their mathematical experiences, lead to misconceptions every bit as powerful, and just as robust, as those in naive physics.

There is a large body of evidence that people's judgments about probability and statistics are strongly affected by their perceptions of everyday experiences. Kahneman and Tversky (1982) have explored what they describe as the "representativeness heuristic": People will assign high probabilities to events that strike them as being prototypical and lower probabilities to events that do not. Kahneman and Tversky provided their subjects with a description of a young woman who was single, outspoken, very bright, and deeply concerned about social issues, and had been a philosophy major in college. They asked a large sample of statistically naive undergraduates to say which was more likely:

A. Linda is a bank teller,
B. Linda is a bank teller who is active in the feminist movement.

Eighty-six percent of the sample chose B over A, despite the fact that since B is a subset of A, the probability of B must be less than the probability of A. Likewise, 50% of a sample of psychology graduate students, who certainly "knew" enough formal mathematics to know better, chose B over A.

Another common statistical misconception concerns what might be called the "law of small numbers," the assumption that independent of sample size, the mean of a sample should be the mean of the population. In one experiment, Pollatsek, Lima, and Well (1981) told students that the average math SAT score in a large school district was 400. If five students were randomly selected from the district and the SAT math scores of the first four were 380, 420, 400, and 600, what did the students expect the fifth score to be? The expected value of any individual score in the population is 400. However, many students assumed that the sample of size 5 should also have a mean of 400. They solved for the "missing score," claiming that the fifth score had to be 200 in order for the sample mean to be correct. As in the case of naive physics, such misconceptions are quite robust. Shaughnessy (1985) reports having presented the same problem to college senior mathematics majors in the middle of a course on probability and statistics, and replicating the results Pollatsek and his colleagues had obtained with nonmajors.

Not all misconceptions are carried over from the "real world:" Some arise from experience with formal mathematics. One such misconception, abstracted in the same way that misconceptions in naive physics are abstracted, is the following. Suppose that during your entire academic career, every mathematics problem that you were asked was in fact a straightforward exercise designed to test your mastery of a small piece of subject matter. You were expected to solve such problems in just a few minutes: If you did not, it meant that you had not understood the material and the material should be explained to you again. Suppose in addition that this scheme was reinforced in class: Problems were expected to be solved rapidly, and teachers gave you the solutions if you did not produce the answers quickly. Having had that experience over and over again, you might eventually codify it as the following (implicit) rule: When you understand the subject matter, any problem can be solved in 5 minutes or less. The stronger form of this rule is even worse: If you fail to solve a problem in 5 minutes, give up. Unfortunately, this story is not hypothetical: My research indicates that this belief and a number of equally counterproductive beliefs about mathematics are all too common among our students. (See chapter 8 for details.)

SOME CONCLUDING COMMENTS

These comments address the interaction between the two disciplines that share the title of this book, cognitive science and mathematics education. There may appear to be some imbalance between the two, because a large part of our conferences (and consequently the material in this book) consisted of discussions of "what's new" in cognitive science. In fact the discussions themselves were highly interactive, with contributions from all the constituent disciplines. The teachers, for example, came to the conferences with insights about students and knowledge about the students' experiences that no laboratory studies could provide. They also came with a deep sense of what is important and what is peripheral to the students' learning experience. The researchers came with some new ideas and perspectives (and some that were not so new, as Joe Crosswhite reminded us), some hard evidence to back them up, and many of questions. Both groups profited from the give-and-take that occurred.

Two examples of that give-and-take stand out in my memory. In the first example, one of the researchers at the conference was describing an anomaly in his data. He could explain most of the data he had gathered, but he was having trouble with one pattern of results. The work from one set of students he had examined demonstrated a consistent sequence of error patterns that differed from those he had seen previously. Not only was that behavior unusual, but the theoretical analyses done by the researcher indicated that such behavior was unlikely. One of the teachers suggested that the students' behavior might have been the result of what they had been taught the previous year. A particular textbook series teaches an antecedent procedure in a nonstandard way, and misinterpretations of that procedure might have led the students to produce the kind of behavior the researcher saw in the laboratory. In this way, the interaction between researchers and teachers allowed a teacher to solve a problem that had puzzled a researcher.

In the second example, which took place during a conversation about students' bugs, research pointed the way to the solution of a teacher's problem. We had been talking about systematic errors in adding and subtracting whole numbers. In the middle of the conversation, one of the teachers realized why her students made a particular mistake when subtracting fractions. Often students need to "borrow" from the whole number part of a fixed fraction when subtracting one mixed fraction from another, for example, when computing

$$7^{24}/_{40} - 5^{35}/_{40}.$$

The teacher recalled that many of her students, in working this problem, had converted

$$7^{24}/_{40} \text{ into } 6^{124}/_{40}.$$

This conversion had always seemed odd, but now it made sense. The students had incorrectly generalized a procedure from whole number subtraction. They had "borrowed" a 1 from the 7 and moved it into the next "column," thus converting the 24 in the numerator of the fraction to 124. (See chapter 7 for a more extended discussion.) To sum up briefly, the "buggy metaphor" provided an explanation of classroom behavior that had been quite puzzling. This time the interaction between research and practice gave a teacher the tools to solve a problem of her own, when she adapted an idea drawn from research. Such interactions among all four constituencies typified the exchanges at the conferences.

Some of those exchanges are recorded in this text. At least two of the papers contain questions posed by teachers for researchers, and the dialogues that follow the papers contain comments from all four groups. The effects of other exchanges were more subtle, but it should be recalled that all of the papers were group projects, and all of them reflect input from mathematicians, mathematics educators, teachers, and cognitive scientists. Our apparent focus on new ideas from cognitive science stems from our desire to get that information out in the open.

Once such information is available, how does one make use of it? The details of cognitive research may not be of interest to anyone except cognitive researchers, although the implications and the underlying ideas may be. How might one exploit them? The answer I propose might be expected from a constructivist: The power of the information lies in what you make of it. For example, consider the teacher's insight about her students' "fraction bug" that was just described. That insight did not come from applying results that were "handed down" from the research. Rather, it came from a teacher's adopting a perspective substantiated by the research—namely, that consistent mistakes made by students are often the result of systematic misapplications or misgeneralizations of procedures that the students have learned—and examining her experience in the light of that perspective. Speaking more broadly, I claim that adopting the constructivist perspective allows teachers (myself included) to see what takes place in their classroom quite differently: When you prepare for a class, the key issue is how students will interpret what you show them rather than how much of it they will absorb. Once you adopt this perspective, the way you teach is altered: among other changes, your job begins rather than ends with the clear presentation of subject matter.

These remarks may suggest that successful teaching involves a lot more work than we had thought, but that is not so. In fact, the constructivist perspective may make teaching easier. With the absorption model of instruction, a teacher had little choice but to repeat the presentation of subject matter (or variants of it) until the lesson had been absorbed. As we know all too well, there were limits to the success of this approach. With the interpretation model, one has the opportunity to "get inside students' heads" and locate the causes of their difficulties—which makes it much easier to remedy them.

A similar argument can be made for any of the research ideas discussed in this book. As an example consider a main theme of this chapter, cognitive process analyses. Few people are likely to make detailed process analyses at the level that Newell and Simon discuss in *Human Problem Solving,* but that is beside the point. The fact is that if you get into the habit of asking questions like How do I go about solving this kind of problem? and What does it really take to make this strategy work? you begin to see the subject matter, and the task of teaching, somewhat differently. If, for example, you reflect on your solutions to mixed sets of algebraic equations, you will discover that you don't solve them in the way textbooks suggest; most likely you use strategies similar to those uncovered by Bundy and his colleagues. Discussing your solution methods with your students—discussing *why* you choose the approaches to mixed problems that you do, and comparing the methods you chose with the methods they chose—may be tremendously useful to them. Susan Carey's comment bears repeating: Human beings are theory builders. If we look both at our own thinking and that of our students from that point of view, there is the potential for significant progress.

Acknowledgments

I would like to thank Andy diSessa, Bat-Sheva Eylon, Jim Greeno, Lisa Quinn, and Fred Reif for their comments on an earlier draft of this chapter.

REFERENCES

Brown, J. S., & Burton, R. R. (1978). Diagnostic models for procedural bugs in basic mathematical skills. *Cognitive Science, 2*, 155–192.

Bundy, Alan. (1975). *Analysing mathematical proofs (or reading between the lines).* (Res. Rep. No. 2). Edinburgh: University of Edinburgh, Department of Artificial Intelligence.

Caramazza, A. McCloskey, M., & Green, B. (1981). Naive beliefs in "sophisticated" subjects: Misconceptions and trajectories of objects. *Cognition, 9*, 117–123.

Carey, S. (1985). *Conceptual change in childhood.* Cambridge, MA: MIT Press.

Carpenter, T. P., Lindquist, M. M., Matthews, W., & Silver, E. A. (1983). Results of the Third NAEP Mathematics Assessment: Secondary school. *Mathematics Teacher, 76*(a), 652–659.

diSessa, A. A. (1983). Phenomenology and the evolution of intuition. In D. Gentner & A. Stevens (Eds.), *Mental models* (pp. 15–33). Hillsdale, NJ: Lawrence Erlbaum Associates.

diSessa, A. (1985, June). *Knowledge in pieces.* Paper presented at 5th annual symposium of the Jean Piaget Society, "Constructivism in the computer age," Philadelphia. Manuscript available from author, School of Education, University of California, Berkeley, CA 94720.

Gagné, R. M., & Briggs, L. J. (1979). *Principles of Instructional design (2nd ed.).* New York: Holt, Reinhart, & Winston.

Ginsburg, H., and Opper, S. (1969). *Piaget's theory of intellectual development.* Englewood Cliffs, NJ: Prentice-Hall.

Gardner, Howard. (1985). *The mind's new science.* New York: Basic Books.

Kahneman, D., & Tversky, A. (1982). On the study of statistical intuitions. In D. Kahnemann, A. Tversky, & P. Slovic (Eds.), *Judgment under uncertainty: Heuristics and biases* (pp. 493–508). Cambridge, England: Cambridge University Press.

Kilpatrick, J. (1978). Variables and methodologies in research on problem solving. In L. Hatfield (Ed.), *Mathematical problem solving.* Columbus, Ohio: ERIC.

Krulik, S. (Ed.). (1980). *Problem solving in school mathematics* (1980 Yearbook of the National Council of Teachers of Mathematics.). Reston, VA: National Council of Teachers of Mathematics.

McCloskey, M. (1983). Naive theories of motion. In D. Gentner & A. Stevens (Eds.), *Mental models* (pp. 299–324). Hillsdale, NJ: Lawrence Erlbaum Associates.

McCloskey, M., A. Caramazza, & B. Green. (1980). Curvilinear motion in the absence of external forces: Naive beliefs about the motions of objects. *Science, 210*, 1139–1141.

McCorduck, P. (1979). *Machines who think.* San Francisco. W. H. Freeman Co.

Meyer. R. A. (1980) Mathematical problem-solving performance and intellectual abilities of fourth-grade children. In J. G. Harvey & T. A. Romberg (Eds.) *Problem solving studies in mathematics.* (pp. 179–198). Madison: Wisconsin Research and Development Center Monograph Series.

National Council of Supervisors of Mathematics. (1977). *Position paper on basic mathematical skills.* Washington, DC: National Institute of Education. Also published in *Arithmetic Teacher,* 25 (October 1977), 19–22.

National Council of Teachers of Mathematics. (1980). *An agenda for action.* Reston, VA: NCTM.

Newell, A., & Simon, H. (1972). *Human problem solving.* Englewood Cliffs, NJ: Prentice-Hall.

Piaget, J. (1954). *The construction of reality in the child.* New York: Basic Books.

Pollatsek, A., Lima S., & Well, A. (1981). Concept or computation: Students' understanding of the mean. *Educational Studies in Mathematics, 12,* 191–204.

Pólya, G. (1957). *How to solve it.* (2nd ed.). Princeton, NJ: Princeton University Press. (Original work published 1945).

Pólya, G. (1981). *Mathematical discovery* (combined paperback edition). New York: WIley.

Resnick, L. B. (1983). Toward a cognitive theory of instruction. In S. Paris, G. M. Olson, & H. W. Stevenson (Eds.), *Learning and motivation in the classroom* (pp. 5–38). Hillsdale, NJ: Lawrence Erlbaum Associates.

Rissland, E. L. (1985). Artificial intelligence and the learning of mathematics: A tutorial sampling. In E. A. Silver (Ed.), *Teaching and learning mathematical problem solving: Multiple research perspectives* (pp. 147–176). Hillsdale, NJ: Lawrence Erlbaum Associates.

Romberg, T. A., & Carpenter, T. P. (1986). Research on teaching and learning mathematics: Two disciplines of scientific inquiry. In M. C. Wittrock (Ed.), *Handbook of research on teaching (3rd ed.).* New York: Macmillan.

Russell, B., & Whitehead, A. (1925–27). *Principia mathematica.* 2nd Ed. Cambridge: Cambridge University Press.

Schoenfeld, A. H. (1985). *Mathematical problem solving.* Orlando, FL: Academic Press.

Shaughnessy, J. M. (1985). Problem-solving derailers: the influence of misconceptions on problem-solving performance. In E. A. Silver (Ed.), *Teaching and learning mathematical problem solving: Multiple research perspectives* (pp. 399–415). Hillsdale, NJ: Lawrence Erlbaum Associates.

Skinner, B. F. (1958). Teaching machines. *Science, 128,* 969–977.

Thorndike, E. L. (1922). *The psychology of arithmetic.* New York: Macmillan.

Wertheimer, M. (1959). *Productive thinking.* New York: Harper & Row.

2

Foundations of Cognitive Theory and Research for Mathematics Problem-Solving Instruction*

Edward A. Silver
Department of Mathematical Sciences
San Diego State University

In recent years, mathematics educators have become more interested in increasing students' ability to use and apply the mathematical knowledge learned in school for solving problems both within and outside of the school setting. The initial emphasis in mathematics was on "problem solving," which included the solution of non-routine problems as well as that of routine textbook exercises. More recently, as the emphasis in problem solving has gained wide acceptance, educators have begun to stress the solution of "real-world" problems, that is, problems with some relevance to the lives of adults or students and with solutions involving the use of some mathematical knowledge.

In the past two decades, the interest of practitioners in problem solving and real-world problems has been supplemented by a considerable amount of research on the learning of mathematics and the use of mathematical knowledge to solve problems. Much of the research has been conducted by cognitive psychologists, who seek to develop and validate theories of human learning and problem solving, and mathematics educators, who seek to understand the nature of the cognitive interaction between students and the mathematical subject matter they study and the problems they solve. This paper summarizes briefly some of the most salient features of the cognitive theory and research and draws from that research a few suggestions for designing mathematics instruction that will

---

*An earlier version of this paper was presented at the annual meeting of the American Educational Research Association, Chicago, IL, April 1985.

produce students who are better equipped than they are now to use their mathematical knowledge to solve problems.

This chapter is divided into three parts. The first section considers some general components of cognitive theory, particularly two basic aspects: memory and information processing. The discussion of memory involves both its contents (e.g., propositions, procedures, episodes, images) and its architecture (e.g., sensory buffer, long-term memory, and working memory); the discussion of information processing deals with the limitations imposed by working memory and ways to overcome those limitations through "chunking" and automatic information processing. (Although this section serves as background for the rest of the paper, the reader who is well versed in cognitive theory can skip to the next one.)

The second part of the chapter presents some specific details concerning problem-solving theory and research. Some of the most important theoretical constructs (e.g., schemas, metacognition) are discussed, along with important related research findings.

The final part of the paper briefly treats some aspects of current cognitive theories of learning. In particular, both the importance of extensive experience with examples and a constructivist view of learning are discussed. Following this discussion of learning theory, a few instructional suggestions are presented. The instructional suggestions, though at a fairly general level, are consistent with the cognitive theory and research discussed earlier.

FOUNDATIONS OF COGNITIVE THEORY

Modern cognitive theories typically rest on assumptions about memory and information processing. As the term information processing implies, they have been heavily influenced by computer metaphors. Indeed, the following description of the basic elements of cognitive theories contains many analogies between human cognitive activity and the information-processing operations of a computer.

Memory

Memory is basic to most modern cognitive theories. In this section, we treat two aspects that are central to cognitive theories of human thought and problem solving: memory contents and memory architecture.

Memory Contents

Memory research has dealt most often with semantic memory, often with a focus on prose test comprehension (cf. Gagné, 1978) rather than on mathematical comprehension or competence. In discussing the contents of semantic memory, researchers have typically distinguished between propositional and procedural representations. Some have also proposed episodic and imagery representations, which they have distinguished from the representations of semantic memory.

Propositional and Procedural Representations. Although theorists and commentators may differ in their terminology, most modern theories of semantic memory distinguish between propositional and procedural knowledge. Anderson (1976) characterized the two types of knowledge as "declarative" and "procedural"; Greeno (1973) called them "propositional" and "algorithmic", and Gagné and White (1978) called them "propositions" and "intellectual skills." Regardless of the terminology used, the distinction is being made between "knowing that" and "knowing how" (Ryle, 1949).

Many modern theories of semantic memory have arisen in connection with the study of language and comprehension, and it is not surprising that these theories assume that information is stored principally in the form of propositions. The popular view is that these propositions are connected in a network structure and that recall largely involves a process of recognizing stored patterns. Thus, semantic memory consists of concepts and relations among concepts; propositions are linked in memory by associations of concepts (or even structures representing concept relations).

Although important for the development of a sound theory of memory, the characterization of basic elements of semantic memory and the nature of the relationship between these basic elements is not critical for our discussion. It is sufficient for us to acknowledge that with respect to mathematics in particular semantic propositional knowledge exists and can be contrasted with procedural knowledge. Considerably less attention has been given to the storage of procedural information (Greeno, 1973), although some interesting work has been done (Anderson, 1976; Gagné & White, 1978). It is often very useful, for the purposes of discussion, to view semantic memory as being partitioned into these two components, but the distinctions are often blurred when one analyzes cognitive activity. As Greeno (1973) has suggested: "Some algorithmic structures are strongly embedded in the structure of propositional knowledge, and others are not. It is for those algorithms that are strongly embedded in propositional knowledge that the distinction between algorithmic and propositional knowledge is particularly dubious" (p. 116). For a further discussion of the relationship between propositional and procedural knowledge in mathematical settings, see Silver (1986).

Episodic Representation. One of the most important distinctions that has been made in memory representations is between semantic memory and episodic memory. The original distinction was made by Tulving (1972), who described semantic memory as "a mental thesaurus, organized knowledge a person possesses about words, and other verbal symbols, their meaning and referents, about relations among them, and about rules, formulas, and algorithms for the manipulation of these symbols, concepts, and relations" (p. 386). In short, semantic memory is memory for symbolic systems. On the other hand, episodic memory "receives and stores information about temporally dated episodes or events and temporal-spatial relations among these events; episodic information is always stored in terms of its autobiographical reference to the already existing contents of the episodic memory store" (pp. 385–386).

Somewhat different characterizations of a semantic–episodic distinction have been proposed by Kintsch (1974) and by Schank (1975). These variations may be the consequence of different interpretations of the terms "semantic memory." In any case, Tulving's original distinction is sufficient for our purposes here.

Semantic and episodic memories are thus seen as different types of representations. Nonetheless, they almost certainly interact with each other. For example, after listening to a lecture on a mathematical topic, one may have forgotten the exact words used by the speaker at specific points, but almost certainly not one's interpretation of the lecture. Both kinds of memory come into play here. The interpretation of the lecture is clearly based on associated concepts stored in semantic memory. An episodic memory of the lecture would not be stored in semantic memory, but semantic memory would probably be used in forming the episodic representation. Episodic memory also influences semantic memory. For example, Kintsch has raised the important theoretical question of how general knowledge about the world develops from particular experiences; presumably the general knowledge is stored as semantic memories, whereas the particular experiences are stored as episodic memories.

Imagery Representations.   The fourth proposed form of memory representation consists of images. In brief, images are representations (usually visual, although auditory or haptic images are certainly possible) that are based on sensory impressions of actual objects or events. Images may also be generated from verbal cues corresponding to the objects or events. Although there is some debate about whether images constitute a form of memory representation or a mode of processing and retrieval, considerable recent research suggests that imagery is an important aspect of human memory.

Imagery has a long history in psychology. Aristotle (1912) championed it as an essential component of human thinking. Although the rise of behaviorism reduced interest in imagery, the past two decades have seen it given considerable research attention. Early studies focused on individual differences to a great extent, but modern work explores other aspects of imagery. The classic finding in this area is that there do not exist people who think in "pure" imagery (Carey, 1915) but that some people appear to have a memory processing style in which images are "dominant" (Pear, 1924). In general, distribution curves from imagery studies indicate that the general population is approximately normally distributed with respect to the use of imagery.

According to Reese (1977), studies of imagery have been based on at least three general characteristics of images, all of which can affect memory: (a) imagery is highly memorable; (b) images are coded distinctly from verbal memory code and thus provide a second mode of access to stored information; and (c) images provide an effective framework for organizing material to be remembered. Assumption (a) has high face validity and was confirmed in several studies reviewed by Reese. Paivio (1971) presented an extensive analysis of assumption (b), the dual-coding notion: that is, the relative independence of verbal memory and imagery. In this area, most of the experimental work has involved memorizing word lists. The possible role of imagery as an organizational factor, assumption (c), is discussed further in connection with manifestations of mathematical abilities.

Memory Architecture

In keeping with the human–machine analogy, most cognitive theorists distinguish among at least three kinds of memory registers: a sensory buffer, or iconic memory; a long-term memory; and a working memory. Figure 2.1 depicts all three as well as the way information flows among them.

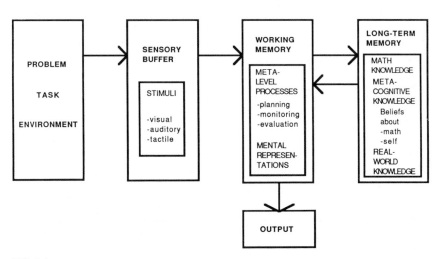

FIG. 2.1.

Sensory Buffer

The sensory buffer, or iconic memory, receives and briefly maintains a visual, auditory, or tactile stimulus. The amount of time the stimulus is maintained is sufficient for it to be recognized, classified, stored in working memory, or ignored. Although the sensory buffer can register a great deal of information simultaneously, it can hold it only briefly. It then either transfers the input to working memory or loses it. Transferring the input to working memory makes it accessible for storage in long-term memory or for interaction with long-term memory's contents.

Long-term Memory

Long-term memory (LTM) is a repository of permanent knowledge and skills. Unlike the sensory buffer or working memory, it seems to have a limitless storage capacity. Although there are many different theories or models of LTM, most theorists agree that the vast amount of information in LTM is organized and structured in some fashion. They typically hypothesize that LTM has a nodal structure, in which the nodes are interrelated in complex ways both within and across networks. A node represents an item of information, or a cluster, or "chunk," of related items. Activating some of the elements of a cluster is likely to activate all of them. Since LTM is the storehouse of all knowledge that a person has about the world, some nodes must contain sensori-perceptual or sensori-motor knowledge and others, cognitive knowledge. In many of the theories, information is organized into conceptual networks, represented by nodes (representing concepts) and lines connecting nodes (representing meaningful associations between concepts). LTM probably contains thousands of these interrelated networks, and the interrelationships among them allow the derivation of information other than that which was explicitly stored (Bower, 1978).

Working Memory

Working memory (Feigenbaum, 1970), or short-term memory, is the place where most of the cognitive "action" takes place. It is here that elements transferred from the sensory buffer are processed for storage in LTM. It is also in working memory that the elements transferred from the sensory buffer can interact with items retrieved from LTM.

Working memory maintains an internal representation of the current state of cognitive activity. Its capacity, however, is limited to a few items of information—generally no more than six or seven. This may sharply limit the size of problems one can deal with successfully (Miller, 1956). In many cognitive theories, this limitation on working memory is of central importance for the development of the cognitive information-processing strategies that are discussed in the next section.

Information Processing

As we have seen, working memory contains the information that is actively being used at any given time. Information can be stored in LTM only after being processed in working memory, and it can be used in thinking only after being retrieved from LTM and placed in working memory. Thus, the cognitive activity called "information processing" consists of controlling the flow of information into and out of working memory by processes such as receiving information from the sensory buffer and retrieving information from LTM; recognizing, comparing, and manipulating symbols in working memory; and storing information in LTM.

The constraints on human thought that are imposed by the limitations on the capacity of working memory are substantial and have dramatic consequences. When the capacity of working memory has been filled, new information that is received into working memory will replace older information. Older information can be updated and kept available through such strategies as rehearsal (verbatim repetition of the information). Nevertheless, these strategies do not increase the capacity of working memory; they merely increase the probability that the information will be held in working memory for a longer period of time and that it will be available when needed.

Efficient methods of processing information can preserve precious storage capacity and increase the amount of information that can be stored. Two such information-processing mechanisms are "chunking" and "automatic processing".

Chunking

Chunks are collections of related items of information represented by a single symbol or concept. For example, the mnemonic device ROY G. BIV represents the colors of the natural spectrum (red, orange, yellow, blue, indigo, violet), and a single item, ROY G. BIV, replaces seven discrete elements in working memory. In mathematics, familiar words and phrases such as "Please Excuse My Dear Aunt Sally" (for the order of operations—parentheses, exponents, multiplication and division, addition and subtraction), FOIL (for the multiplication of terms in the product of two binomials), and SOHCAHTOA (for the basic trigonometric relations) are all examples of chunks commonly used for remembering important information.

Chunking does not apply only to mnemonic devices but also to any assemblage of discrete bits of information into a whole that can be stored more easily in memory. People often use chunking to remember such things as their social security number, a familiar pattern of pieces on a chess board (deGroot, 1965), or even a set of formulas for solving problems in mechanics (Chi, Feltovich, & Glaser, 1981). Chunking helps make it possible to process a great deal of detailed information automatically, as do automatic information-processing procedures.

Automatic Processing

Some theorists (e.g., Schneider & Shiffrin, 1977) have argued that there are two kinds of information processing: controlled and automatic. In a controlled process, a sequence of nodes is activated by a controlling subject. Because a controlled process requires attention and most cognitive models assume serial (rather than parallel) processing, only one sequence can operate at a time. An automatic process also involves activation of a set of nodes, but under the control of a particular input to working memory (internal or sensory) rather than under the subject's control. The sequences are carried out automatically without the subject's attention. Thus, automatic processes use little or none of the capacity of working memory, and training or practice may develop a high degree of automaticity.

Many of the behaviors engaged in while driving an automobile or riding a bicycle as involve automatic processing. In mathematics, much of the instruction given in arithmetic algorithms appears to be directed at automatizing the procedures of numerical computation that start out as controlled processes.

One common difficulty that students have when solving complex or novel mathematics problems is the apparently overwhelming number of processes to control. Students' problem-solving abilities might improve greatly if they could use working memory more efficiently, that is, if they learned to use automatic processing for the more routine elements of an activity, and thus made resources available for the controlled processing of the novel aspects of solving the assigned problems.

PROBLEM SOLVING:
THEORY AND RESEARCH

In this section we turn our attention to some aspects of cognitive theory and research that are directly relevant to problem solving. As we noted at the outset, problem solving involves the application of one's knowledge to tasks that may be well structured or poorly structured, familiar or unfamiliar, simple or complex. Thus, if we are interested in people's ability to use their knowledge of mathematics, then the problem-solving literature is the most natural place to look for information and guidance.

In the first part of this section we examine some important aspects of current theories of problem solving. In particular, we discuss the components of problem-solving models and certain fundamental processes hypothesized to constitute problem-solving activity. We then turn our attention to some of the salient findings from research on mathematical problem solving. This part of the paper focuses specifically on the information about successful problem solving that is provided in the literature.

Theories of Problem Solving

Most current theories of problem solving are based on information-processing assumptions of the sort described in the first part of this chapter. For example, in Newell and Simon's (1972) seminal theory the problem-solving process involves the rapid serial processing of information, the results of which are held in working memory (possibly in the form of chunks to compensate for working memory limitations) and brought into contact with other information retrieved from LTM. The search for relevant and useful information in LTM is facilitated by the organization of LTM information in networks.

Most information-processing descriptions or models of problem solving contain three major components: (a) problem task environment, (b) long-term memory, and (c) working memory. Inasmuch as we have already discussed the last two items in some detail in the first part of the chapter, their treatment here will be brief and restricted to their role in models of problem solving. The process of mental representation in problem solving—another theme of central importance or cognitive theories—is also treated in this section.

Problem Task Environment

The task environment is the structure of facts, concepts, and the interrelationships among them that constitute the problem. In many mathematical settings, the problem task environment may serve two distinct functions: (a) allowing access to external information and (b) providing an external memory for information generated by the problem solver. External information includes the statement of the problem itself, any directions and goals, and auxiliary information such as tables, notes, sample problems, or related text materials. The external memory makes available for examination partial solutions, facts from long-term memory useful in solving the problem, and notes concerning the process of solving the specific problem. As a problem solution progresses, a problem solver may alter sources of external information and may re-evaluate the salience of the information.

Long-Term Memory

Long-term memory contains mathematical knowledge, such as basic facts, processes, generalized problem types, heuristics, and algorithms. It contains beliefs and opinions about mathematics, about one's self as a learner or doer of mathematics, and other metacognitive knowledge. It also contains knowledge about the real world, knowledge about the quantities involved in a problem, and other knowledge that may be related to the problem setting. Information from long-term memory may be accessed and used in the working memory; or it may be accessed, held briefly in the working memory, and then placed into the external memory storage in the problem task environment.

Working Memory

The most active component is working memory. It is here that the information obtained from the external problem task environment inter-acts with the knowledge retrieved from LTM, and here that the metalevel processes of planning, monitoring, and evaluating take place, if at all. Working memory is also the site of the process of mental representation.

Mental Representation

In the original Newell and Simon (1972) formulation, the idea of a *problem space*, the problem solver's mental representation of the task environment, was an important component of the problem-solving model. Current formulations use the term "representation" to describe the problem solver's internalized version of the problem task.

Current theories, however, do not specify the nature of mental repre-sentations completely. The term applies both to the current depiction of a problem or task in working memory and to the permanent depiction of knowledge in LTM: We say that the problem is *represented* in working memory, and that we retrieve knowledge *representations* from LTM. Moreover, the problem task itself has various representations associated with it, including at least the mathematics that it represents and the forms of representation (e.g., natural language, graph, equation) used in its description. In a fairly complete exegesis of the term representation, Kaput (1985) presents an interesting discussion of some of the issues involved.

In most current formulations of human problem solving, the differences and interactions between the external task environment and the internalized problem space, or problem representation, are seen as critical to the solution of a problem. Two situations illustrate how these differences and interactions come into play. A problem may have a simple and straightforward task environment, but the form of its presentation may make it difficult for the solver to develop an adequate problem representation. In other problems, it may be necessary to use information that is not contained in the problem task environment, such as information drawn from assumptions implicit in the problem or knowledge about terms and relationships that the problem statement assumes is available to the problem solver.

A popular view of the problem-solving process holds that solvers construct an initial mental representation of the problem while reading the problem or shortly afterwards. This representation then changes as it interacts with further information from the task environment or with knowledge retrieved from LTM, resulting in the construction of a more elaborated representation of the problem. The problem is solved when a completely adequate representation has been obtained.

The quality of a solver's problem representation is thus seen as central to the problem-solving process. An inaccurate or incomplete problem representation may make solving a problem difficult or impossible, and the quality of the solution obtained depends on the adequacy of the problem representation.

SUCCESSFUL PROBLEM SOLVING:
RESEARCH PERSPECTIVES

A considerable amount of research has been done in the area of problem solving. Although it is not possible to present a complete summary of the themes and findings of this vast literature, this part of the paper presents some trends. In particular, we focus on the nature of successful problem solving and organize the discussion around five themes: pattern recognition, representation, understanding, memory schemas, and meta-processes.

Pattern Recognition

In a classic study, deGroot (1965) asked chess experts (grandmasters and masters) and novices (ordinary chess players) to reproduce the position of the pieces on a chess board. The pieces were either arranged in a midgame position or randomly arranged on the board. For the midgame positions, the experts were able to reproduce the positions of the 20 or 25 pieces almost without error, whereas ordinary players could place only a half dozen pieces correctly. For the random arrangements, only about six pieces were correctly placed both by masters and ordinary players.

The ability of chess experts to recognize patterns of related pieces on the board almost instantaneously, and to use these patterns rather than the positions of individual pieces in processing information, is consistent with the findings of other research on expert knowledge in complex task domains. Skilled medical diagnosis may also involve pattern-recognition skills. In one study (Norman, Jacoby, Feightner, & Campbell, 1979), written case histories were presented to doctors with varying levels of experience and training (practicing physicians, third-year residents, first-year residents, and second-year medical students). The subjects were asked to read each case history and then write out as much of it as they could remember. For the histories based on common diseases, experienced physicians recalled the most details, followed by the other groups in descending order of experience and training. For the histories that contained findings not suggestive of any disease, there was little difference among the groups. Skilled diagnosis apparently involves perceiving patterns of signs and symptoms that correspond to disease entities.

It is possible that pattern recognition skill may involve episodic or imagistic memory representations that are built up through many hours or years of experience and unlikely that rapid pattern-recognition skill can develop without extensive practice in the task domain.

Representation

As we noted earlier, many theories describe problem solving as a process of building successively richer and more refined problem representations. The solver begins with an initial representation, then gradually elaborates and refines it until he or she obtains a final problem representation that is adequate for the solution. Lesh (1985) proposes an alternative view in which the problem solver builds and then abandons unstable representational models of the problem until reaching a stable model. In either view, problem representations are central to the problem-solving process. Consequently, investigators have looked at the representations formed by successful problem solvers to see if they are in any way different from those formed by less successful solvers.

Larkin (1980) studied the problem solving of experts and novices in the area of physics. She noted that experts frequently formed qualitative representations of the problems before attempting any quantitative analyses. They often mentally replaced the original problem with an abstracted version that retained its general structure and features and then used this idealized representation as a guide in solving the original problem. In contrast to the experts, novices usually initiated quantitative analyses even when their problem representations were inadequate and insufficiently constrained to suggest the correct procedures.

Clement (1983) has noted that experts in mathematics and some scientific domains engage in metaphorical processes as they construct problem representations: that is, they look for analogies between the problem at hand and other situations with which they are familiar, and they use these analogies to suggest possible representations of the problem to be solved. Clement notes that these analogies often take the form of mental images.

Representations play an essential role in the problem solver's understanding of the problem and in the recognition of relationships with other problems. Students who build similar representations for mathematically related problems are far more likely to notice their similarity and to use the relationship about one problem in solving the other. Conversely, the similarity between mathematics problems with isomorphic representations can go unnoticed if the solver does not represent the two problems in similar ways.

Understanding

The first stage of Pólya's (1957) descriptive model of problem solving is: "understand the problem." Understanding the problem at the outset of a problem-solving episode is roughly synonymous with constructing an initial problem representation. There is considerable evidence suggesting that failure to solve problems can often be attributed to failure to understand the problem adequately: that is, failure to construct adequate initial problem representations.

Unsuccessful problem solvers are often characterized by their general lack of problem understanding. As we have noted, novices in a task domain often proceed to solve problems without constructing an adequate initial problem representation, that is, without understanding the problem. Nevertheless, there is a more sanguine view of naive problem solving that may also be extracted from the research literature. This more optimistic view bears directly on the description of problem-solving success.

There is evidence that without instruction children can invent correct procedures for the performance of a number of tasks. For example, without explicit instruction, many children develop correct mental arithmetic strategies. Moreover, many are able to solve simple addition and subtraction story problems without any formal instruction (Carpenter, 1985). Analysis of children's successful solution processes suggests that they attend to the semantics of the problem situation; that is, they succeed because they have an adequate understanding of the problem situation. As Carpenter notes:

> Even before they have received instruction in formal arithmetic, almost all children exhibit reasonably sophisticated and appropriate problem-solving skills in solving simple word problems. They attend to the content of the problem; they model the problem; they invent more efficient procedures for computing the answer. Given the limits of their mathematical knowledge, this performance is remarkable. (p. 37)

This finding is consistent with other research on problem solving that supports the importance of semantic understanding. It is striking that many of these same children will no longer attend to the semantics of the problem after receiving mathematics instruction in school; they prefer instead to choose an arithmetical operation on the basis of the problem's surface features.

Memory Schemata

Simon (1980) has noted that "research on cognitive skills has taught us . . . that there is no such thing as expertness without knowledge—extensive and accessible knowledge." Pólya (1973) observed earlier that "a well-stocked and well-organized body of knowledge is an asset to the problem solver. Good organization which renders the knowledge readily available may be even more important than the extent of the knowledge." As we have seen, information stored in LTM plays an important role in problem-solving theories. Because the efficient retrieval of information may depend on how information is organized in LTM, differences in problem-solving success may be partly attributable to differences in problem solvers' knowledge organization.

The idea of a memory schema (a cluster of knowledge that describes the typical properties of the concept it represents) has recently helped explain many aspects of human knowledge organization and recall, especially in the area of prose text learning. In the past 5 years or so, much research has been generated on the influence and use of schemata. A schema is usually described (e.g., Thorndyke & Yekovich, 1980) as representing a prototypical abstraction of a complex and frequently encountered concept or phenomenon, and it is usually derived from past experience with numerous exemplars of the concept involved. Schemata have been associated with not only interpreting and encoding incoming information, but also recalling previously processed information (Thorndyke & Hayes-Roth, 1979). They can influence the efficiency with which information is recalled from memory (e.g., Mandler & Johnson, 1977).

Furthermore, schemata may account for inferences drawn from incomplete information (e.g., Bransford, Barclay, & Franks, 1972).

One prominent example of the role played by schemata in successful problem solving comes from a study that examined the differences between experts and novices in physics (Chi, Feltovich, & Glasser, 1981). Both groups were asked to categorize problems according to similarities in solution methods. The novices tended to sort on the basis of surface features, such as inclined planes, pulleys, and friction, whereas the experts categorized problems on the basis of the fundamental principles of physics that were involved, such as conservation of energy and Newton's third law. The greater knowledge and experience of the experts made it possible for them to represent problems in terms of schemata containing both factual knowledge and procedural knowledge—schemata that at times included the relevant formulas for solving those kinds of problems.

Silver (1979) found that successful problem solvers were far more likely than unsuccessful ones to relate and categorize mathematics problems on the basis of their underlying similarities in mathematical structure. Unsuccessful problem solvers were more likely to rely on surface similarities in problem setting or context, or the question asked in the probem, in judging problem similarity.

The first explicit link between schema theory and mathematical problem solving was made by Hinsley, Hayes, and Simon (1977). In their series of studies, subjects were instructed to (a) categorize a series of standard algebra word problems into groups of related problems; (b) categorize a given problem after hearing only a portion of the problem text; (c) solve some standard algebra word problems and some nonstandard problems, in which the cover stories and underlying structures did not match; (d) solve two "nonsense problems" that were constructed by taking standard algebra problems and replacing some of the content words with nonsense words; and (e) solve the "small town" problem, a standard distance-rate-time problem to which some irrelevant information concerning the right triangular relation of three problem elements was added.

Hinsley et al. (1977) reported that subjects sorted the problems into standard categories, such as distance-rate-time or age problems, and that subjects were able to categorize a problem almost immediately, usually after hearing only the first few words. For example, after hearing the three words, "A river steamer . . ." one subject said, "It's going to be one of those river things with upstream, downstream, and still water. You are going to compare times upstream and downstream—or if the time is constant, it will be the distance." Furthermore, they reported that subjects tended to use problem categorizations to assist in retrieving useful solution information from long-term memory. In the problems that were susceptible to multiple interpretation, subjects who catergorized the problems as distance-rate-time problems attended to different information in the problem statement than did subjects who categorized it as a Pythagorean triangle problem. Hinsley et al. concluded that in general, their subjects did have schemata for standard algebra problems and that the schemata influenced the encoding and retrieval of information during problem solving.

Mayer (1982) conducted a series of experiments in which subjects read a set of standard algebra word problems and then were asked to recall each one and to construct problems based on certain situations (e.g., "trains leaving stations"). He found that subjects recalled relevant information much better than irrelevant details, recalled high frequency problem forms (i.e., common standard problem types) better than low frequency forms, made recall errors that converted low frequency forms to high frequency forms, and constructed problems that matched standard textbook forms. He interpreted his findings as supporting the hypothesis that students possess schemas for standard algebra problems and that the schemas guide the encoding and retrieval of problem information.

It is also possible to develop schema-based interpretations of several common problem-solving phenomena. For example, Loftus and Suppes' (1972) finding that a word problem in a sequence of problems is more difficult to solve if it is preceded by problems of a different type and Luchins' (1942) observations of *einstellung,* or mental set, can both be explained in terms of difficulties encountered in shifting from one schema to another. Furthermore, there is a straightforward schema-based explanation of a student who has difficulty solving any problem that deviates from a standard textbook form: Locked into a particular approach, the student lacks the flexibility to adapt to new circumstances.

Robert Davis and his colleagues (Davis, 1984; Davis, Jockusch, & McKnight, 1978) have written extensively about schemata and the related notion of a "frame." They have demonstrated that these two concepts may explain many aspects of algebraic task performance, including, but not limited to, the solution of algebra word problems. Here too the fundamental intuition reflects a generalized pattern recognition. Triggered by the recognition of certain features of a problem, the student selects a solution method—whether or not ot "fits" the problem situation.

Schemata are useful not only for retrieving clusters of related and useful information, but also for shaping the representation of problems. In research dealing with young children's ability to solve arithmetic and subtraction story problems, for example, the differences between highly skilled and less skilled performance by children has been modeled in terms of the development of more powerful problem schemata for representing the problems. (e.g., Briars & Larkin, 1984; Riley, Greeno, & Heller, 1981)

Meta-processes

The cognitive science perspective suggests that extensive domain-specific knowledge appears to be vital to success in problem solving. It is reasonable to ask what other kinds of knowledge are involved in skillful problem solving. Several cognitive researchers interested in mathematical problem solving (e.g., Lester & Garofalo, 1982; Schoenfeld, 1985; Silver, 1985) have argued for more attention to metacognitive aspects of the problem-solving process. These processes—such as initial assessment of personal competence or problem difficulty or managerial decisions regarding allocation of cognitive resources—often appear to be the "driving forces" of a problem solving episode. According to Flavell (1979), meta-cognition refers to one's knowledge of the cognitive processes and products of oneself and others. It also refers to the self-monitoring, regulation, and evaluation of cognitive activity.

Metacognition is not a new construct; provision for metacognitive functioning has been made in most information-processing models of cognition. Atkinson and Shiffrin (1968) used the term "control processes" to identify certain voluntary and strategic behaviors that help one remember, like rehearsing a telephone number or tying a string around one's finger. Butterfield and Belmont (1975) posited the existence of an "executive function" that selects appropriate control processes on the basis of task and environmental constraints. One of the earliest models of limited metacognitive functioning and regulation was provided in the TOTE (Test/Operate/Test/Exit) units proposed by Miller, Galanter, and Pribam (1960). In Skemp's (1979) model of intelligence, metacognitive processes are contained in the domain and functions of delta-two. And Case (1974), in his description of neo-Piagetian approaches to information processing, has included an "executive scheme" that is primarily concerned with plans and regulation. The terms control processes, executive function, reflective intelligence, and executive scheme are not synonymous with metacognition, but they all cover, to a great extent, the area in which metacognition is operative.

Although many models of human information processing include meta-cognitive processes, there has been relatively little empirical investigation of them. Indeed several important questions immediately suggest themselves: questions regarding the accessibility of metacognitive processes for systematic study, the role of "awareness" in the acquisition or utilization of metacognitive behaviors, the locus of observation points from which to view metacognitive aspects of problem solving, and the nature of the relationship between metacognitive skills and problem-solving performance. In addition, questions regarding the nature of the monitoring and evaluating processes need to be carefully formulated and investigated. It is clear that these processes make important contributions to human problem solving; it is far more difficult to develop a formal descriptive model that accounts for the role that these processes play.

Also of interest is the role that *beliefs* about mathematics or about problem solving play in skillful problem solving. Although Norman (1981) has identified "belief systems" as one of the important issues for cognitive science and Carbonell (1981) has constructed a mechanism that incorporates knowledge contained in belief systems (in this case, political ideologies) into the process of formulating plans for action, researchers working on problem solving have tended to ignore the role of beliefs in skillful problem solving. It seems clear that no process model of problem solving in any domain can be complete without an adequate account of the role of metacognition and belief systems.

Some progress has been made in explicating certain metacognitive processes, such as planning. For example, Sacerdoti (1977) has developed a fairly detailed theory of planning in terms of the organization of knowledge about actions into a procedural network. Each action has a set of preconditions, a set of consequences, and a set of subactions that are necessary for its completion. This type of organization facilitates planning, because it allows one to begin with the larger units and then proceed to the subactions. Since planning depends on knowledge, the ways in which planning occurs probably depend on the problem solver's knowledge of the domain. For the person experienced in the domain, planning may be automatic, whereas the novice must generate and try out various sequences. Given the importance of metaprocesses, much more theoretical and empirical work needs to done on the metacognitive aspects of problem solving.

LEARNING THEORY AND INSTRUCTIONAL RECOMMENDATIONS

In this section we consider briefly some theoretical perspectives on learning taken by cognitive scientists. In particular, we consider both the constructivist view of learning and the notion that the development of expertise depends heavily on extensive experience with examples within a particular task domain. Following our consideration of learning theory, we turn our attention briefly to some instructional recommendations that are consistent with the work discussed in this paper.

Learning Theory

Recent years have seen mathematics educators taking a greater interest in the work of cognitive scientists. This is due, at least in part, to the rediscovery of learning as an important topic in psychology. Although some researchers engaged in the early phases of Artificial Intelligence work were apparently very interested in learning, it is only in recent years that much attention has been given to creating systems that learn. Perhaps in reaction to the obvious interest of behaviorist psychologists in learning (associationist bonds), cognitive psychologists ignored learning in most of their early work, choosing to focus almost exclusively on issues of performance. Recently, however, the cognitive science community has begun to show great enthusiasm for studying the mechanisms of learning.

Those of us in mathematics education with a particular interest in instruction should welcome this change, for understanding how mathematics is learned will help us design more effective instruction. David Wheeler (1981) has aptly characterized the peculiar connection between teaching and learning:

> Suppose that instead of taking the position that teaching should provoke or cause learning we try saying that *effective teaching is contingent upon learning* . . . [This position] strongly suggests that teachers and educators would have to *know* learning in a way that at present they do not: that they would have to become students of learning, not merely practitioners of it. (p. 2)

Expertise Depends on Extensive Experience

In this section we briefly discuss two themes that are well represented in the cognitive science literature on learning: (a) expertise is derived from extensive experience with examples and (b) learning is a constructive process. These themes are consistent with, and have evolved from the research characterizing the expert/novice differences in various task domains.

The first theme is that expertise develops only after extensive experience in the domain of expertise. A corallary to this theme is that expertise develops over an extended period of time. The design of systems that learn has taken both the theme and its corollary into account. For example, Larkin (1980) developed a program called ABLE which learned to solve increasingly complex (though fairly elementary) physics problems by using its problem-solving experiences to augment its store of knowledge. ABLE's knowledge consists of "productions" (condition-action pairs). It utilized this knowledge to solve problems by matching the condition part of a production with the contents of the program called

"Barely ABLE." The program learned to improve its performance by acquiring new productions as a result of its problem-solving experiences with representative problems. Moreover, Larkin's simulations of novices (Barely ABLE) and experts (ABLE) were quite similar in many respects to the behavior of human problem solvers.

Rissland (1978, 1985) has stressed the fundamental role of examples in the teaching/learning process. She suggests a classification scheme for examples and their functions: (a) start-up examples (easily understood and presented cases), (b) reference examples (standard cases), (c) counter-examples (falsifying cases), (d) model examples (paradigmatic cases), and (e) anomolous examples (exceptions and pathologies).

The importance of examples is also evident in other analyses of the development of expertise. In their summary of how expertise is acquired, Chase and Chi (1980) emphasize not only the role of examples but also the length of time needed to develop expertise. They ask how expertise develops and respond that

> the most obvious answer is practice, thousands of hours of practice . . . There may be some as yet undiscovered basic abilities that underlie the attainment of truly exceptional performance, . . . but for the most part practice is by far the best predictor of performance. . . . Practice can produce two kinds of knowledge . . . a storage of patterns of lexicons [and] a set of strategies (or procedures) that can operate on the patterns. (p. 12)

They suggest that expertise develops when extensive experience with a rich set of examples creates a highly textured knowledge base. They also argue that there appears to be no limit, except for physiological dysfunction, to the extent to which domain-specific cognitive skills can be developed.

The Constructive Nature of Learning

The second theme is that human learning is largely a constructive process. John Anderson and his colleagues have been among the most active cognitive learning theorists. Anderson (1982; Neves & Anderson, 1981) proposes a detailed theoretical account of the acquisition of problem-solving expertise that involves three stages: (a) a *declarative* stage, when the learner receives instruction that is encoded only as a set of facts about the skill; (b) a *knowledge compilation* stage, when the knowledge is converted into a set of procedures that can be carried out without any interpretive operations; and (c) a *procedural stage,* when the activity can be carried out autonomously.

One of the central ideas in Anderson's (1982) theory that is also shared by many other cognitive theorists is the constructive nature of the learning process:

> One of the fundamental assumptions of cognitive learning psychology is that new knowledge is in large part "constructed" by the learner. Learners do not simply add new information to their store of knowledge. Instead, they must connect the new information to already established knowledge structures and construct new relationships among those structures. This process of building new relationships is essential to learning. It means that mathematical knowledge—both the procedural knowledge of how to carry out mathematical manipulations and the conceptual knowledge of mathematical concepts and relationships—is always at least partly "invented" by each individual learner." (pp. 249–250)

An interesting consequence of the constructive nature of the learning process is that "bugs"—minor systematic flaws in otherwise correct procedures—can occur. Brown and his colleagues (Brown & Burton, 1978; Brown & VanLehn, 1980) have extensively studied the bugs that arise in children's learning of the subtraction algorithm in elementary school. As VanLehn (1986) has recently argued, the extensive data on subtraction bugs not only supports a constructivist view of children's learning, but also suggests that the traditional instructional model of "teacher tells student what to do" is not likely to be consistently effective.

Cognitive scientists have captured two central features of human learning in their theoretical and research accounts. Nevertheless, much work still needs to be done in learning theory. I have not yet seen a compelling cognitive science account of the processes of abstraction, generalization, or cognitive reorganization, for example. Furthermore, the view of learning that is shared by many cognitive scientists—that learning is an accretion of productions—is not consistent with the conservatism of human learning (i.e., the ubiquitous tendency of humans to try to "get by" with available knowledge), nor does it account for the apparent discontinuities that are observed as human learners transform themselves from novices to experts. It remains to be seen whether adaptive production systems, or other systems that learn, will account for these important characteristics of human learning. Given the substantial progress that has been made already, there is reason to be optimistic.

INSTRUCTIONAL RECOMMENDATIONS

In this section we consider a few recommendations for instruction that can be drawn from the theoretical and empirical foundations just discussed. But first, a few caveats! These are recommendations, not implications or logically necessary consequences. Many of them could be made for other reasons or could emerge from other types of analyses. Moreover, the current limitations of a cognitive theory of learning may mean that the foundation upon which the recommendations rest is not as solid as it should be. Most of the suggestions made in this section should, however, ring true to practitioners. Finally, I have striven to strike a balance between the specificity that some people seek in instructional suggestions and the vague generalities that fill the literature. I do not, for example, make specific recommendations for teaching a particular skill or cluster of skills, such as a multidigit subtraction algorithm or the solution of algebra story problems, although some theory-based research might support them. Rather, I have tried to give general recommendations that can be implemented in a variety of specific instances in the mathematics curriculum. For more specific suggestions, the interested reader should seek out the studies cited in the earlier sections of this chapter and in other chapters in this volume.

As we have seen, the cognitive science view of learning is colored to a great extent by the empirical study of expert behavior in complex task domains, or by comparisons of expert and novice behavior in those domains. Although precollege mathematics education aims primarily at helping the average student attain reasonable proficiency in a wide variety of mathematical task domains, the theoretical and empirical perspective of expert–novice comparisons can be valuable to those of us seeking instructional recommendations. In particular, the views that learning is a constructive process and that expertise develops through extensive experience with examples together form a foundation for the recommendations made in this section.

Situational Problem Solving

The first instructional recommendation assumes that instructional time is limited. We should maximize opportunities for learning and the development of proficiency by choosing examples carefully, recognizing the constructive nature of learning. Given the well-documented difficulty that most students have in applying mathematical knowledge to solving problems, the curriculum and the teacher should provide students with rich situational problem-solving opportunities. One way of facilitating the association between situations with applicable mathematical knowledge might be to use prototypical problem situations for introducing and developing instruction on mathematical concepts and skills. If this is done, then it should be easier to apply that knowledge to similar situations encountered at a later time—for the learner's mental representation of the mathematical knowledge would probably consist of a well-connected propositional and procedural network composed of both the particular concepts and skills and rich connections between the elements of that knowledge and the prototypical problem situations to which the knowledge is applicable.

It is possible, however, that situation-based instruction might lead to knowledge that is limited, context bound, and context dependent. To avoid that unfortunate consequence, instruction should build toward general structures that include not only the prototypical examples considered but also problem situations that have not been considered. Moreover, as Rissland's (1985) analysis suggests, it is also important to consider cases for which the procedure or concept is not applicable, such as counterexamples or pathologies.

Focus on Metalevel Processes

It is generally acknowledged that metalevel processes, such as planning, monitoring, and evaluation, are important components of mathematical problem-solving behavior. Nevertheless, these processes are seldom made the explicit focus of instruction. If our goal is to help students to become effective problem solvers, then our instruction must address these important metalevel processes.

Students may need explicit instruction in monitoring and evaluating their problem-solving behavior. One technique that has been found to be successful is having students solve problems in pairs, in which one student acts as the "solver" and the other acts as the "monitor." The role of the monitor is to ask questions that clarify the nature of the problem-solving activity of the solver. After solving one or more problems in this way, the students can exchange roles. Students at the secondary school and college levels have found this to be a useful technique for exploring the role and importance of monitoring and evaluation processes in problem solving.

To emphasize the role of planning, students might be asked to design an explicit plan for the solution of a problem before beginning any calculations. In many cases this would require the students to develop a *qualitative* rather than a *quantitative* approach to the problem at hand. Since qualitative problem analyses early in the solution of a problem are characteristic of expert behavior in mathematics, it is possible that students might increase their problem-solving proficiency as they become more skilled at planning.

If a major goal of mathematics instruction is the development of students' problem-solving abilities, metalevel processes need to become an important curricular focus. Until these processes receive explicit attention in the curriculum, we will continue to produce mathematics students who know fairly well what to do in routine and simple problem situations, but have little competence in handling unfamiliar or complex problems.

Model Problem-solving Processes

Inasmuch as students learn extensively from and by examples and the teacher is the students' primary example of a mathematics problem solver, it should be productive for teachers to model explicitly problem-solving processes in the classroom. Textbooks provide a rich supply of examples from which a learner can generate an understanding of the conceptual or procedural knowledge being studied. Moreover, textbooks can, and often do, provide examples that have been worked out specifically to demonstrate applications of the procedural and conceptual knowledge. However, textbooks almost never provide examples of knowledge representation, or of heuristic or metalevel problem-solving processes. Exemplifying these processes is, therefore, up to the teacher.

Explicit knowledge representation processes, such as creating symbolic, pictorial, diagrammatic, or graphical models of a problem situation, need to be exemplified and discussed by mathematics teachers. Students need to see that there is often more than one way to represent the information in a problem and that correspondences can be made between and among the various representations. As students gain skill in external knowledge representation, it is likely that they will not only become more skillful in handling memory overload in complex problem solving, but also develop greater facility with the mental representation of mathematical information.

Teachers need to model for students the processes of selection and implementation of useful heuristic processes, such as drawing diagrams or examining extreme cases or analyzing a simpler problem. To incorporate these processes into their problem-solving repetoire, students need to see the processes utilized and hear them explained. The demonstration and modeling needs to focus not only on *what* is being done but also on *why* the choice was made.

One way of modeling heuristic and metalevel processes involves teachers' systematically acting out problem solving episodes. Teachers might regularly pose nontrivial problems to be solved collectively and lead the metalevel strategic discussion. They might also pose problems to be solved individually or in groups and precede the work session with a metalevel discussion. Small group cooperative problem solving could be especially conducive to fostering metalevel discussions and problem analyses. (See chapter 8 of this volume for an extensive discussion of these and related issues.)

TEACH TO THE "HIDDEN" CURRICULUM

As a result of their experiences in mathematics classrooms, students develop a set of beliefs about mathematics and about mathematical problem solving. According to Schoenfeld (1985) and the results of the third National Assessment of Educational Progress (Carpenter, Lindquist, Matthews, & Silver, 1983; NAEP, 1983), the majority of junior high and secondary school students believe that mathematics is mostly memorization, that there is usually one right way to solve every mathematics problem, and that mathematics problems should be solved, if at all, in a few minutes or less.

These statements reflect part of what I would label the "hidden" mathematics curriculum. These statements, and others like them, reflect students' beliefs about and attitudes toward mathematics. The students' beliefs and attitudes have been shaped by their school mathematics experiences. Despite the fact that neither the authors of the mathematics curriculum nor the teachers who taught the courses had intentional curricular objectives related to students' attitudes toward and beliefs about mathematics, students emerged from their experience with the curriculum and the instruction with these attitudes and beliefs. Since the students' viewpoint represented by these statements is clearly inadequate, and potentially harmful to their future progress in mathematics, we need to focus our attention more clearly on those hidden products of the mathematics curriculum.

In designing mathematics curricula or planning instructional activities, we need to be mindful that students will integrate their experience with the activity, unit, or course that we are preparing with their prior experiences to form or to modify attitudes toward and beliefs about mathematics and mathematical problem solving. Let us teach toward this hidden curriculum to allow our students to develop attitudes and beliefs that reflect a view of mathematics as vibrant, challenging, creative, interesting, and constructive. Our students may realize greater educational benefits from our attention to the hidden curriculum of beliefs about and attitudes toward mathematics than from any improvement we could make in the "transparent" curriculum of mathematics facts, procedures, and concepts.

REFERENCES

Anderson, J. R. (1976). *Language, memory and thought*. Hillsdale, NJ: Lawrence Erlbaum Associates.

Anderson, J. R. (1982). Acquisition of cognitive skill. *Psychological Review, 89*, 369–406.

Aristotle (1912). Psychology. In B. Rand (Ed.), *The classical psychologists* (pp. 45–83). Boston: Houghton Mifflin.

Atkinson, R., & Shiffrin, M. (1968). Human memory: A proposed system and its control processes. In G. H. Bower & J. T. Spence (Eds.), *The psychology of learning and motivation: Advances in theory and research* (Vol. 2). New York: Academic Press.

Bower, G. H. (1978). Contacts of cognitive psychology with social learning theory. *Cognitive Therapy and Research, 2,* 123–146.

Bransford, J. D., Barclay, J. R., & Franks, J. J. (1972). Sentence memory: A constructive versus interpretive approach. *Cognitive Psychology, 3,* 193–209.

Briars, D. J., & Larkin (1984). An integrated model of skill in solving elementary word problems. *Cognition and Instruction, 1,* 245–296.

Brown, J. S., & Burton, R. R. (1978). Diagnostic models for procedural bugs in basic mathematical skills. *Cognitive Science, 2,* 155–192.

Brown, J. S., & VanLehn (1980). Repair theory: A generative theory of bugs in procedural skills. *Cognitive Science, 2,* 379–426.

Butterfield, E. C., & Belmont, J. M. (1975). Assessing and improving the executive cognitive functions of mentally retarded people. In I. Bailer & M. Sternlicht (Eds.), *Psychological issues in mentally retarded people.* Chicago: Aldine.

Carbonell, J. (1981). Politics. In R. C. Schank & C. K. Riesbeck (Eds.), *Inside computer understanding: Five programs plus miniatures* (pp. 259–307). Hillsdale, NJ: Lawrence Erlbaum Associates.

Carey, N. (1915). Factors in the mental processes of school children. I. Visual and auditory imagery. *British Journal of Psychology, 7,* 453–490.

Carpenter, T. P. (1985). Learning to add and subtract: An exercise in problem solving. In E. A. Silver (Ed.), *Teaching and learning mathematical problem solving: Multiple research perspectives* (pp. 17–40). Hillsdale, NJ: Lawrence Erlbaum Associates.

Carpenter, T. P., Lindquist, M. M., Matthews, W., & Silver, E. A. (1983). Results of the third NAEP Mathematics Assessment: Secondary school. *Mathematics Teacher, 76,* 652–659.

Case, R. (1974). Structures and strictures: Some functional limitations on the course of cognitive growth. *Cognitive Psychology, 6,* 544–74.

Chase, W. G., & Chi, M. T. H. (1981). Cognitive skill: Implications for spatial skill in large-scale environments. In J. Harvey (Ed.), *Cognition, social behavior, and the environment* (pp. 111–136). Hillsdale, NJ: Lawrence Erlbaum Associates.

Chi, M. T. H., Feltovich, P. J., & Glaser, R. (1981). Categorization and representation of physics problems by experts and novices. *Cognitive Science, 5,* 121–152.

Clement, J. (1983, April). *Analogical problem solving in science and mathematics.* Paper presented at the Annual meeting of the American Educational Research Association, Montreal, Canada.

Davis, R. B. (1984). *Learning mathematics.* Norwood, NJ: Ablex.

Davis, R. B., Jockusch, E. & McKnight, C. C. (1978). Cognitive processes in learning algebra. *Journal of Children's Mathematical Behavior, 2,* 10–320.

deGroot, A. D. (1965). *Thought and choice in chess.* The Hague: Mouton.

Feigenbaum, E. A. (1970). Information processing and memory. In D. A. Norman (Ed.), *Models of human memory.* New York: Academic Press.

Flavell, J. H. (1979). Metacognition and cognitive monitoring: A new area of cognitive-developmental inquiry. *American Psychologist, 34,* 906–11.

Gagné, E. D. (1978). Long-term retention of information following learning from prose. *Review of Educational Research, 48,* 629–665.

Gagné, R. M., & White, R. T. (1978). Memory structures and learning outcomes. *Review of Educational Research, 48,* 187–222.

Greeno, J. G. (1973). The structure of memory and the process of solving problems. In R. Solso (Ed.), *Contemporary issues in cognitive psychology. The Loyola symposium* (pp. 103–133). Washington, DC: Winston.

Hinsley, D. A., Hayes, J. R., & Simon, H. A. (1977). From words to equations—meaning and representation in algebra word problems. In M. Just & P. Carpenter (Eds.), *Cognitive processes in comprehension* (pp. 89–106). Hillsdale, NJ: Lawrence Erlbaum Associates.

Kaput, J. (1985). Representation and problem solving: Methodological issues related to modeling. In E. A. Silver (Ed.), *Teaching and learning mathematical problem solving: Multiple research perspectives* (pp. 381–398). Hillsdale, NJ: Lawrence Erlbaum Associates.

Kintsch, W. (1974). *The representation of meaning in memory.* Hillsdale, NJ: Lawrence Erlbaum Associates.

Larkin, J. H. (1980, October). *Models of skilled and less skilled problem solving in physics.* Paper presented to the NIE-LRDC Conference on Thinking and Learning Skills, Pittsburg.

Lesh, R. (1985). Conceptual analyses of problem-solving performance. In E. A. Silver (Ed.), *Teaching and learning mathematical problem solving: Multiple research perspectives* (pp. 309–330). Hillsdale, NJ: Lawrence Erlbaum Associates.

Lester, F., & Garofalo, J. (1982). *Mathematical problem solving: Issues in research.* Philadelphia: Franklin Institute Press.

Loftus, E. F., & Suppes, P. (1972). Structural variables that determine problem-solving difficulty in computer-assisted instruction. *Journal of Educational Psychology, 63,* 521–42.

Luchins, A. S. (1942). Mechanization in problem solving. *Psychological Monographs, 54* (6, Whole No. 248).

Mandler, J. M. & Johnson, N. S. (1977). Remembrance of things parsed: Story structure and recall. *Cognitive Psychology, 9,* 111–151.

Mayer, R. E. (1982). Memory for algebra story problems. *Journal of Educational Psychology, 74,* 199–216.

Miller, G. A. (1956). The magical number seven, plus or minus two: Some limits on our capacity for processing information. *Psychological Review, 63,* 81–97.

Miller, G. A., Galanter, E., & Pribram, K. (1960). *Plans and the structure of behavior.* New York: Holt, Rinehart, and Winston.

National Assessment of Educational Progress (1983). *The third national mathematics assessment: Results, trends and issues.* Denver, CO: Educational Commission of the States.

Neves, D. M., & Anderson, J. R. (1981). Knowledge compilation: Mechanisms for the automatization of cognitive skills. In J. R. Anderson (Ed.), *Cognitive skills and their acquisition* (pp. 57–84). Hillsdale, NJ: Lawrence Erlbaum Associates.

Newell, A., & Simon, H. A. (1972). *Human problem solving.* Englewood Cliffs, NJ: Prentice-Hall.

Norman, D. A. (1981). Twelve issues for cognitive science. In D. A. Norman (Ed.), *Perspective on cognitive science* (pp. 265–295). Norwood, NJ: Ablex.

Norman, G. R., Jacoby, L. L., Feightner, J. W., & Campbell, E. J. M. (1979). Clinical experience and the structure of memory. *Proceedings of the Eighteenth Annual Conference on Research in Medical Education.* Washington, DC: Association of Medical Colleges.

Paivio, A. (1971). *Imagery and verbal processes.* New York: Holt, Rinehart & Winston.

Pear, T. H. (1924). Imagery and mentality. *British Journal of Psychology, 14,* 291–299.

Pólya, G. (1957). *How to solve it* (2nd ed.). New York: Doubleday.

Pólya, G. (1973). *Induction and analogy in mathematics*. Princeton, NJ: Princeton University Press.

Reese, H. W. (1977). Imagery and associative memory. In R. V. Kail & J. W. Hagen (Eds.), *Perspectives on the development of memory and cognition* (pp. 113–175). Hillsdale, NJ: Lawrence Erlbaum Associates.

Riley, M. S., Greeno, J. G., & Heller, J. I. (1981). Development of word problem solving ability. In H. P. Ginsberg (Ed.), *Development of mathematical thinking* (pp. 153–196). New York: Academic Press.

Rissland, E. L. (1978). Understanding understanding mathematics. *Cognitive Science, 2*, 4, 361–383.

Rissland, E. L. (1985). Artificial intelligence and the learning of mathematics: A tutorial sampling. In E. A. Silver (Ed), *Teaching and learning mathematical problem solving: Multiple research perspectives* (pp. 147–176). Hillsdale, NJ: Lawrence Erlbaum Associates.

Ryle, G. (1949). *The concept of mind*. London: Hutchinson.

Sacerdoti, E. G. (1977). *A structure for plans and behavior*. New York: Elsevier.

Schank, R. C. (1975). The structure of episodes in memory. In D. G. Bobrow & A. Collins (Eds.), *Representation and understanding* (pp. 237–272). New York: Academic Press.

Schneider, W., & Shiffrin, R. M. (1977). Controlled and automatic human information processing: I. Detection, search, and attention. *Psychological Review, 84*, 1–66.

Schoenfeld, A. H. (1985). Metacognitive and epistemological issues in mathematical understanding. In E. A. Silver (Ed.), *Teaching and learning mathematical problem solving: Multiple research prespectives* (pp. 361–380). Hillsdale, NJ: Lawrence Erlbaum Associates.

Silver, E. A. (1979). Student perceptions of relatedness among mathematical verbal problems. *Journal for Research in Mathematics Education, 10*, 195–210.

Silver, E. A. (1985). Research on teaching mathematical problem solving: Some underrepresented themes and needed directions. In E. A. Silver (Ed.), *Teaching and learning mathematical problem solving: Multiple research perspectives* (pp. 247–266). Hillsdale, NJ: Lawrence Erlbaum Associates.

Silver, E. A. (1986). Using conceptual and procedural knowledge: A focus on relationships. In J. Hiebert (Ed.), *Conceptual and procedural knowledge: The case of mathematics*. Hillsdale, NJ: Lawrence Erlbaum Associates.

Simon, H. A. (1980). Problem solving and education. In D. T. Tuma & R. Reif (Eds.), *Problem solving and education: Issues in teaching and research* (pp. 81–96). Hillsdale, NJ: Lawrence Erlbaum Associates.

Skemp, R. R. (1979). *Intelligence, learning, and action*. New York: Wiley.

Thorndyke, P. W., & Hayes-Roth, B. (1979). The use of schemata in the acquisition and transfer of knowledge. *Cognitive Psychology, 11*, 82–106.

Thorndyke, P. W., & Yekovich (1980). A critique of schema-based theories of human story memory. *Poetics, 9*, 23–49.

Tulving, E. (1972). Episodic and semantic memory. In E. Tulving & W. Donaldson (Eds.), *Learning strategies* (pp. 383–408). New York: Academic Press.

VanLehn, K. (1986). Arithmetic procedures are induced from examples. In J. Hiebert (Ed.), *Conceptual and procedural knowledge: The case of mathematics*. Hillsdale, NJ: Lawrence Erlbaum Associates.

Wheeler, D. (1981). Editorial. *For the Learning of Mathematics, 1*, 2.

3

Instructional Representations Based on Research about Understanding

James G. Greeno
University of California, Berkeley

COGNITIVE ANALYSES PROVIDE EXPLICIT HYPOTHESES ABOUT TACIT KNOWLEDGE

Cognitive research often analyzes information structures and processes that students need to succeed in instructional tasks, and expresses these conclusions as models in the form of computer programs that simulate successful performance of the tasks. Examples of these include models of reading comprehension (e.g., Thibadeau, Just, & Carpenter, 1982; van Dijk & Kintsch, 1983), models of solving problems in mathematics (e.g., Brown & Burton, 1980; Greeno, 1978), and models of solving problems in physics (e.g., Larkin, McDermott, Simon, & Simon, 1980).

When a computational model is developed for some instructional task, we frequently discover performance requirements that were not evident when we merely thought about the task. This happens because our objective is that the model actually run and produce solutions to problems, answers to questions, or whatever the performance of the task involves. Constructing a computer program is analogous to working through a complete proof of a theorem in mathematics: it ensures that the conclusion that seems to follow from particular assumptions really does follow—that is, it demonstrates the sufficiency of the model's assumptions. When constructing a model reveals that assumptions previously considered sufficient are insufficient, we gain new understanding of the knowledge and processes that are required for student success in the instructional tasks, especially when developing the model reveals gaps we can fill.

These discoveries tell us, at least hypothetically, about *tacit knowledge,* knowledge needed for performing a task, but whose presence is unsuspected by the performer. Polanyi (1959) distinguished between explicit knowledge, "what is usually described as knowledge, as set out in written words or maps, or mathematical formulae," and tacit knowledge, "such as we have of something we are in the act of doing" (p. 12). A cognitive model of a task has to include hypotheses about the properties of tacit knowledge as well as about the explicit knowledge necessary to perform the task. The model, then, is an explicit hypothesis about both explicit and tacit knowledge.

Analyses of tacit knowledge have important implications for instruction. Explicit knowledge corresponds directly to things we can discuss and observe. We can ask a question to find out whether a student knows a multiplication fact or that disjoint sets have no members in common. But tacit knowledge is, by definition, knowledge that is not "set out in written words or maps, or mathematical formulae"—we do not know how to display it directly. Therefore, it can only be communicated implicitly, as an unseen and unanalyzed component of performance. It exists "between the lines" (Bundy, 1975) in examples that are presented in texts or by teachers; and if students are to acquire it, they have to "read between the lines" and do so by a strongly abductive process. Hence, the difficulty of teaching tacit knowledge.

Some investigators have designed resources for instruction that present explicit representations of information that is usually implicit. These representations take information that is generally embedded in cognitive procedures and present that information explicitly. Other investigators provide a way for the student to construct representations. Such explicit expressions of tacit knowledge can enable improved instruction in two ways.

First, a larger proportion of students might succeed. If students can now acquire important components of a skill only by an abductive process that includes filling in crucial "missing pieces" of knowledge, then the process of acquiring those component skills may contribute significantly to their difficulty in learning the whole skill. Providing explicit expressions of tacit knowledge could demystify the tasks significantly and thus enable teachers and texts to communicate more completely with students about what is to be learned. This approach expands upon a principle of instructional design that is familiar in behavioral task analyses and has a tradition that extends at least back to Thorndike (e.g., 1922) and that has recently been developed in more detail by Gagné and his associates (e.g., Gagné & Briggs, 1979). The idea is that instruction should be given for the components of a complex skill, as well as for the integrated form of the skill. Students may be unable to acquire a complex combination of knowledge units if they have not already acquired the units that have to be combined. Cognitive theory extends this idea by providing analyses of a different set of components—cognitive processes and knowledge structures—in addition to the behavioral components that are considered in the earlier theories.

Explicit hypotheses about tacit knowledge could also be a resource for strengthening students' metacognitive understanding and skill. (See chapter 8 for a discussion of metacognition.) Students who succeed in tasks are not aware of the processes that depend on tacit knowledge, but an explicit hypothesis about those processes provides a medium for reflection and discussion about the processes. This could increase students' awareness of their own cognitive processes and provide more opportunities to monitor and evaluate their own progress in understanding and learning.

COGNITIVE MODELS OF THE UNDERSTANDING OF PROBLEMS HAVE BEEN DEVELOPED IN RESEARCH

The cognitive process of understanding problems is a topic involving tacit knowledge that has recently received considerable research attention. "Understanding" is a word with many meanings, but in this discussion its sense is quite simple: Understanding is assumed here to be a process that takes information from a situation and converts it into a form that is used in a cognitive process.

It is widely recognized that understanding is an important part of problem-solving success. Until recently, we have had little or no explicit knowledge about understanding; that is, the knowledge involved in understanding problems has been almost entirely tacit. Elementary mathematics texts emphasize the importance of understanding the problem, but most texts offer little advice for the student other than to read the problem carefully and think about which arithmetic operation to use. Algebra texts often have a unit on translating sentences into equations, but the process of translation assumes knowledge of the meaning of what is translated, and most texts provide little help in characterizing these meanings.

A cognitive model of understanding and solving problems simulates the process of understanding by constructing representations based on the words in problem texts. The representations contain information that students appear to gather from the texts and use in their solutions. The process can be characterized as the recognition of patterns of information.

Elementary arithmetic word problems that involve addition and subtraction have received considerable research attention. Several investigators who were working on these problems, largely unbeknownst to each other, met at a conference in 1979 and discovered that they had independently arrived at similar characterizations. At that time research was focusing on characterizing patterns of information as they are presented in different kinds of word problems, and analyses by Carpenter and Moser (1982), Nesher (1982), Riley, Greeno, and Heller (1983), and Vergnaud (1982) proposed similar sets of patterns.

Riley et al.'s (1983) version emphasizes a distinction in the process of understanding between identifying quantities and recognizing relations among quantities. Identifying a quantity involves recognizing a component of information that specifies a number of some kind of object, an amount of some kind of substance, a rate, or some other quantitative element. Quantities are different from numbers; in fact, numbers are attributes of quantities. In many cases, a quantity is specified in text by a phrase combining a numeral with a term for units, such as "six books," "eight liters," or "55 miles per hour." (For a more general discussion, of quantities and numbers, see Schwartz, 1984.)

With the quantities in a problem identified, the problem solver also recognizes a pattern of relations within the pattern. In Riley et al.'s (1983) version, there are three patterns involving addition and subtraction. First, there may be an event that changes a quantitative value; for example a person may own some books, whose number is either increased (by finding or buying some more books) or decreased (by losing or giving some away.) Second, there may be two quantities that are not changed, but are thought of both as individual quantities and as a combined whole or set, such as two persons, who each own some books, which are considered as a number being owned "altogether." Third, there may be two individual quantities to be compared; for example, two persons each own some books, and a comparison is made of "how many more" one person owns than the other.

Each of these patterns involves three quantities, and the relations can be specified as a set of roles (Kintsch & Greeno, 1985). In a change pattern, one of the quantities is the starting quantity, the second is the amount of change, and the third is the result. In a part–whole pattern, two of the quantities are parts and the third is the whole. In a comparison, one quantity is the base or referent quantity, the second is compared to the referent, and the third is the difference between the other two.

If about six patterns for multiplicative relations are added to the system, the problems used in the elementary grades all can be understood. Many multiplication problems include intensive quantities, that is, quantities that denote a number or amount in each unit of a total quantity. Typical examples are the number of apples in each box or the amount of gasoline used per mile by an automobile (cf. Schwartz, 1976; Kaput,

1985). A common pattern involves an intensive and an extensive quantity; the intensive quantity, the amount in each unit, and the extensive quantity, the number of units, are combined to form a total amount. Another pattern contains two intensive quantities, the amount of *a* per *b*, and the amount of *b* per *c*, which combine to give another intensive quantity, the amount of *a* per *c*. There also are multiplicative comparisons in which one quantity is *a* times another quantity; ratio problems involving four quantities in a proportional system, three of which are usually given; fractional part–whole problems in which a quantity is specified as a fraction of another quantity and in which the quantities thus denote the fractions as well as the amounts in the parts; and combinatorial problems involving the numbers in two sets and the number in their Cartesian product. (Vergnaud, 1983, worked out a somewhat different set of multiplicative patterns.)

Computational models have been developed that simulate the understanding and solution of word problems. Riley et al. (1983) developed a model that constructs representations of quantitative patterns. Using concepts developed by van Dijk and Kintsch (1983), Kintsch and Greeno (1985) considered text processing more systematically. The model they developed was implemented as a computer program by Fletcher (1984) and has been extended to apply to a wider range of problems by Dellarosa (1985).

A discussion of the following example will illustrate the ideas in these models: "Jay had nine books; then he lost five books. Now how many books does he have?" A model is given a set of propositions corresponding to the text:

$$P1 \quad x = Jay$$
$$P2 \quad HAVE(x, P3)$$
$$P3 \quad NINE(BOOKS)$$
$$P4 \quad LOSE(x, P5)$$
$$P5 \quad FIVE(BOOKS)$$
$$P6 \quad HOWMANY(BOOKS)$$
$$P7 \quad HAVE(\times, P6)$$

(Many extant models of parsing are able to convert texts to propositional representations of this general kind.)

The propositions in the text base are used to construct a representation of the three specified quantities: Jay's nine books, the five books he lost, and the unknown number of remaining ones. Each of these is represented using a schema that corresponds to a set with three attributes: the kind of object in the set, a specification of the set (in this case, the owner), and the set's cardinality. Figure 3.1 shows this information in the form of a semantic network for one of the sets in this problem.

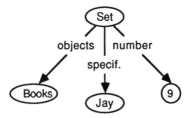

FIG. 3.1.   Semantic network representation of a set.

The representation of a problem consists of several units like Fig. 3.1, arranged in a pattern. For this example, the pattern is called Transfer, and it has three roles: start, transfer-in or transfer-out, and result. The model constructs a representation like the one in Fig. 3.1 for the each of the three quantities as well as the pattern shown as a semantic network in Fig. 3.2.

The model simulates solution of the problem, based on the information in the representation. It has simulated knowledge of problem-solving operators that can be applied to solve many problems, including the one in Fig. 3.2, and it uses procedures that construct and modify sets of blocks corresponding to quantities in the problem. (Riley et al., 1983, considered data from situations in which children of varying ages, including pre-school children, had solved problems with the aid of blocks.) For the problem described in Fig. 3.2, the model simulates the construction of a set of nine blocks, the removal of five, and the discovery of the answer, which involves counting the remaining blocks.

Similar representations are constructed, using the Part–Whole and Comparison schemata, for problems such as: "Jay has three books; Jay and Kay have seven books altogether; how many books does Kay have?" and "Jay has three books; and Kay has seven books; how many more books does Kay have than Jay?" For some problems, the representation that the model constructs does not permit a solution using the available problem-solving operators, and the model of successful performance must elaborate the representation with another schema. The following problem demonstrates the need for such elaboration: "Jay had some books; then he lost five books; now he has four books. How many books did he have in the beginning?" The semantic network that the model constructs is in Fig. 3.3. Lacking problem-solving operators to apply to the initial representation that was constructed with the Transfer schema, the model applies another schema, the Part–Whole schema, to obtain a representation for which its problem-solving operators do apply.

The hypothesis that patterns of this general kind are important in understanding word problems is supported by the results of experiments

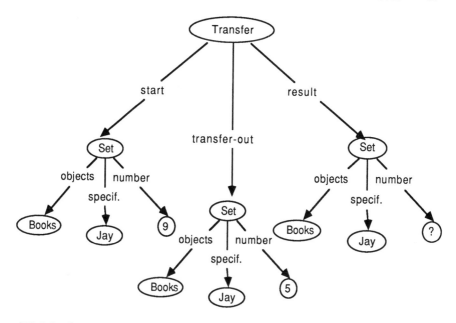

FIG. 3.2.   Semantic network for a problem based on a transfer schema.

in which problems of different kinds were given to students of different ages. We formulated versions of the model that differed by the amounts of knowledge we hypothesized were necessary for successful performance. It turned out that a natural sequence of models could be constructed simulating the performance of children at different ability levels: The simplest models simulate children who solved only a few problems; intermediate models simulate those who solved more; and the most complex models simulate those who solved nearly all. The models predict the overall difficulty of different kinds of problems, as well as specific patterns of correct and incorrect performance followed by individual students, and the data obtained were reasonably consistent with these predictions. (Analyses of specific response patterns are reported in detail by Riley & Greeno, 1985.)

The models developed by Riley et al. (1983) and Kintsch and Greeno (1985) include specific procedures for constructing representations like those of Fig 3.1, 3.2, and 3.3. These representations are based on schemata that are assumed to be included in children's knowledge and that provide general structures. Understanding occurs when the information taken from specific problems is fitted into one of these general structures. Briars and Larkin (1984) developed an alternative model, which has simpler knowledge about information patterns and constructs simpler representations of problem texts, but uses more complex prob-

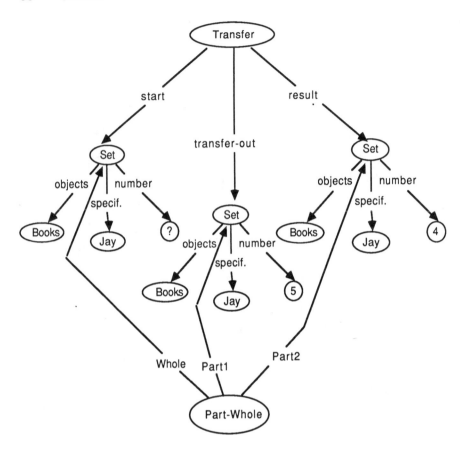

FIG 3.   Semantic network for a problem with Transfer converted to Part-Whole.

lem-solving operators. In Briars and Larkin's (1984) model, relations among quantities like those represented explicitly in the schema-based model are recognized implicitly in the conditions for applying procedures that construct and modify sets.

Briars and Larkin (1984) obtained evidence supporting their model by analyzing the cognitive requirements of different kinds of problems. According to their model, some problems dealing with particular sets of objects require more complex information structures than others dealing with the same sets of objects. In addition, problems differ in how directly they cue construction of sets and inferences about them. Briars and Larkin used data from several instruments to estimate the regression coefficients associated with the factors that should contribute to problem difficulty; these factors accounted for 88% of the variance among problem types. Detailed observations of student efforts to construct and modify sets of objects, and of specific wrong answers given by students, also agreed quite well with predictions of the model.

The schema-based model developed by Riley, Greeno, Kintsch, and their associates and the model by Briars and Larkin, which relies on more complex cognitive procedures, are different in ways that are significant for cognitive theory. They agree in a crucial point: Recognizing patterns that involve quantities and relations among quantities is a major requirement for successful performance in arithmetic word problems. This provides a substantial basis for the design of innovative materials that can help improve the teaching of the processes involved in understanding problems.

INSTRUCTIONAL REPRESENTATIONS OF INFORMATION STRUCTURES OF PROBLEMS HAVE BEEN DESIGNED, USING THE COGNITIVE ANALYSES

The models of understanding word problems illustrate the general point that cognitive analyses provide detailed hypotheses about tacit knowledge, especially when the analyses are formulated in detail as computer programs. While the importance of understanding problem texts has been well recognized, cognitive analyses of the processes and information structures involved in understanding go further: They provide much more definite and specific hypotheses about the patterns of information that students need to recognize in the texts and about the cognitive processes involved in that recognition. These hypotheses pass the test of sufficiency.

These models contain explicit descriptions of knowledge that is ordinarily tacit. In other words, they describe processes of which students and teachers are unaware: Students do not know what they do when they succeed in understanding a word problem, and teachers do not know what students have to do to succeed in these tasks. The models provide hypotheses about those kinds of tacit knowledge.

Once these hypotheses have been developed, questions arise about their applicability to instruction. What would happen if we found ways to present explicit representations of the information patterns that students need to recognize in word problems as part of their instruction? Instructional materials have been designed in an attempt to find out.

Despite that kind of overlap between theoretical cognitive analysis and instructional design, they are very distinct activities. Cognitive theory provides hypotheses about the knowledge and skill of successful student problem solvers and the ways in which their knowledge and skill differ from those of less successful ones. Instructional design attempts to create activities and materials that can help students to acquire the knowledge and skill necessary for success. Although cognitive models can describe more or less accurately the knowledge and skill we want students to acquire, the experiences that will help students acquire that knowledge and skill constitute a separate issue.

Cognitive analyses can, however, offer strong suggestions regarding instructional design, especially when they focus on the properties of tacit knowledge. In the case of word problems, models of the cognitive processes involved in understanding indicate that students need to recognize quantities and relations among quantities. Perhaps presenting explicit representations of the quantities and their relations would enhance instruction, at least for some students.

Here we have a special case of a general principle in instructional design: Providing explicit instruction in the components of the knowledge and skill we want students to acquire can be more helpful than teaching a complex combination whose individual components may be unknown to the students. Because these models describe the processes and information structures involved in understanding problems, they illustrate the capabilities of cognitive theory to characterize instructional objectives different from those of earlier behavioral task analyses.

In this section, I describe two instructional representations that have been developed by investigators at the Learning Research and Development Center for use in teaching arithmetic word problems. The two systems present information at different levels of abstraction. One shows dots in enclosed regions, representing objects and sets, and uses several graphical devices to represent relations between the sets. The other, more abstract representation consists of semantic networks in which the nodes of the network represent quantities, and links between the nodes represent relations by which quantities are composed or decomposed to form other quantities.

A Representation of Objects, Sets, and Relations for Primary-Grade Students

The first system I describe was designed by C. Mauritz Lindvall and Joseph Tamburino. It has been used for experimental instruction with kindergarten and first-grade children.

Lindvall's representations of problems are diagrams that constitute models of the situations described in the problems. Dots corresponding to objects in the problems are contained by closed curves representing sets. Relations between sets are represented in ways that correspond to the different patterns of quantities typical of addition and subtraction word problems. Figure 3.4 shows examples of the diagrams used for eight kinds of problems in an instructional study by Tamburino (1982).

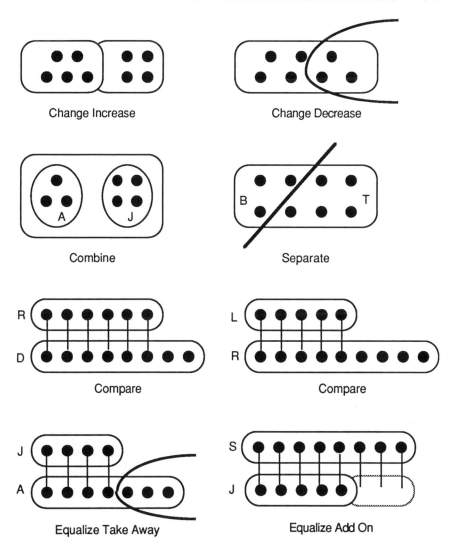

FIG 3.4 Diagrams for illustrative word problems in Lindvall's system (Tamburino, 1982).

Although the diagrams used by Lindvall and Tamburino appear to differ substantially from those used in the Kintsch and Greeno (1985) model, there is a natural correspondence between the problems types that are understood by the models and these instructional representations. The Change Increase pattern represents Transfer-In problems such as "Sue had 5 pencils. She got 4 more pencils. How many did she have then?" The Change Decrease pattern is used for Transfer-Out problems such as "May had 7 cookies. She then ate 3 of them. How many did she have left?"

The Combine and Separate patterns represent problems that are understood with the Part–Whole schema. The Combine pattern represents problems such as "Ann had 3 apples. Jill had 4 apples. How many apples did they have altogether?" The Separate pattern represents problems such as "Together Bob and Tony had 8 toy cars. 3 of these were Bob's. How may did Tony have?"

The Compare pattern is used for problems such as "Rick had 6 kites. Dan had 8 kites. How many more kites did Dan have than Rick?" and "Len had 5 books. Rita had 9 books. How many less books did Len have than Rita?" According to Kintsch and Greeno's (1985) model, these problems are understood with the comparison schema.

The Equalize patterns combine features of comparison and change. They are thus understood with the comparison and transfer schemata. Equalize Take Away is used for problems such as "Jim had 4 cookies. Al had 7 cookies. How many cookies would Al have to eat to have as many as Jim?" Equalize Add On represents problems such as "Sally had 8 rings. Jan had 5 rings. How many more would Jan have to get to have as many as Sally?"

Tamburino (1982) instructed 22 primary-grade students in using diagrams like those in Fig. 3.4 to help solve word problems and tested the instruction's effect on the students' problem-solving ability. The instruction consisted of a series of 22 lessons in which students were taught to construct the eight types of diagrams shown in Fig. 3.4 for word problems. Performance on a 20-item test of word problems similar to those used in instruction improved by an average of 3.4 items. More important, performance on a test of transfer, including problems requiring two steps, problems with larger numbers, and problems involving continuous quantities, improved by an average of 4.2 problems in a 20-item test. These instructional materials also have been used successfully in a school setting with standard classroom presentations (Waller, 1983).

A Representation of Quantities and Relations for Middle-School Children

The second representational system was developed by Valerie Shalin and Nancy Bee (1985a, 1985b). It was implemented as a computer graphics system by Ernest Rees.

The representations in Shalin's system are more abstract than those in Lindvall's and allow representations of more complex problems. In Shalin's representation of a problem, the objects correspond to the quantities in the problem rather than to individual physical objects. For example, for "Ann had 5 pencils," Shalin's representation would have a single graphical object with a label, "Ann's pencils," and the numerical property 5. This representation is shown in Fig. 3.5. Recall that in Lindvall's representation, this would be represented as five dots inside a closed curve.

In the cognitive models of word problem solving that have been developed, understanding a problem includes identifying the quantities in the problem and recognizing the relations among them. Using Shalin's representation system forces the problem solver to be explicit about relations that are otherwise tacit. For example, the system uses a different set of symbols for quantities that will be combined by multiplication (e.g. "number of bags" and "number of books per bag") than it does for quantities that will be combined by addition (e.g. "number of books" and "number of additional books"). The representation system uses a theory of semantic types to classify the quantities in problems. Four types are distinguished: extensive quantities, differences, intensive quantities, and factors. As shown in Fig. 3.6, a different shaped "hat" represents each type.

Rectangular hats, shape (a), are used for extensive quantities, as in "Ann had 5 pencils." Triangular hats, shape (b), are used to represent differences; in "Dan has 2 more kites than Rick," the difference of two between Dan's kites and Rick's kites would be represented with a box of type (b). Rounded hats, shape (c), are used to represent intensive quantities. For example, in "Tom put 6 books in each bag," the number of books per bag is a quantity represented with shape (c). Trapezoidal hats, shape (d) are used for multiplicative factors; in "Sue had 3 times as many pencils as Ann," the factor 3 would be represented with shape (d).

To represent a problem, boxes are used for quantities and links between the boxes for relations. The relations are based on the theory of types,

Ann's pencils

5

FIG 3.5.   Representation of a quantity in Shalin's system.

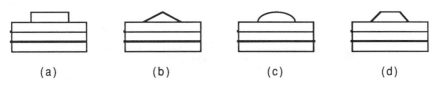

(a)   (b)   (c)   (d)

FIG 3.6   Shapes used to represent different types of quantities in Shalin's system.

which includes rules for the composition of quantities. An example is in Fig. 3.7, which shows the diagram for the simple problem: "Tom had to deliver 48 books, and he put 6 books in each bag. How many bags did he use?"

Of the three quantities, the number of books and the number of bags are extensive, and the number of books per bag is intensive. The three quantities form a compositional unit, as indicated by the lines connecting the boxes. The form of the composition is a version of parts and a whole; in general, the amount per unit combines with the number of units to form the total amount. By convention, the amount per unit and the number of units are placed beside each other, and the total amount is placed above them in the diagram. The dots in the diagram mark the quantities that are given in the problem, and the arrow marks the unknown requested by the problem.

The theory of compositions specifies the quantitative types that can be produced by composition with other types. Table 3.1 shows a set of additive compositions involving extensive quantities and differences. Two extensive quantities can combine to form another extensive quantity and the result is a superset or combined object composed of previously independent subsets or parts. When an extensive quantity and a difference combine, they form another, larger extensive quantity. Finally, two differences can combine to form another difference.

TABLE 3.1
Additive Compositions

	Extensive	Difference
Extensive	Extensive	Extensive
Difference		Difference

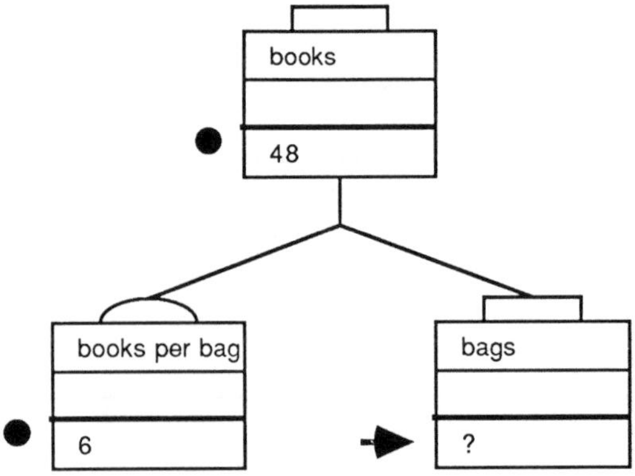

FIG. 3.7.   Representation of a simple problem in Shalin's system.

Table 3.2 shows multiplicative compositions that can be formed. A factor (i.e., *a* times as many or *b* times as much) can be combined with any type of quantity to obtain another one of the same type. The composition of an intensive and an extensive quantity is exemplified by a problem that combines an amount per unit multiplicatively with a number of units to form a total amount—itself an extensive quality. An intensive quantity and a difference produce a difference in the same way.

TABLE 3.2
Multiplicative Compositions

	Extensive	Difference	Intensive	Factor
Extensive	Extensive	Difference	Extensive	Extensive
Difference		φ	Difference	Difference
Intensive			Intensive	Intensive
Factor				Factor

Two intensive quantities form another intensive quantity: that is, *a* per *b* and *b* per *c* produce *a* per *c*. Two extensive quantities can be multiplied to form another extensive quantity, the Cartesian product of the component quantities. A similar composition results from combining an extensive quantity multiplicatively with a difference. That differences do not seem to combine multiplicatively is the one exception to an algebra of multiplicative compositions of quantitative types. There are also some additive compositions involving intensive quantities and factors. (For a more complete discussion of these compositions, see Greeno et al., 1986.)

The theory of semantic types that is reflected in Shalin's system is not sufficient to determine whether quantities can be composed. There are additional constraints that depend on the units of the quantities. For example, to compose an extensive and an intensive quantity multiplicatively, the units of the extensive quantity must match the "denominator" of the units of the intensive quantity, so that there is *a* per *b*, times an amount of *b*, resulting in an amount of *a*. An instructional representation called the Semantic Calculator, which focuses on the units of quantities, has been developed by Schwartz (1982). Shalin's system does not explictly deal with constraints based on units, but implementing the system as an intelligent tutor would require including knowledge that could evaluate compositions according to the units of quantities, as well as type.

The computer implementation of Shalin's representation system includes a facility for adding information that is inferred from the information given in the problem. For example, the diagram shown in Fig. 3.7 can be modified by adding inferred information, as is shown in Fig. 3.8: The answer has replaced the question mark in the unknown quantity, and an expression has been inserted in the box that represents the number of

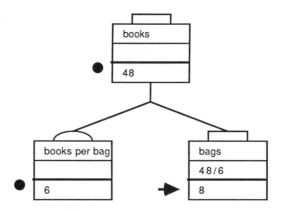

FIG. 3.8.   Representation of the Solution of a simple problem in Shalin's system.

bags and indicates that the numerical value of that quantity is obtained by dividing the total number of books by the number of books per bag, that is, 48 by 6.

Shalin's system also enables the representation of more complex problems. For multiple-step problems, more quantities are given, as there are quantities that are not mentioned in the problem but correspond to intermediate results. Figure 3.9 illustrates the following problem: "The charge for parking is $0.65 per hour. Ms. Jones parked for 4 hours. How much change did she get from a $10 bill?" The amount of money charged for parking four hours is not given or mentioned as an unknown, but it is needed as an intermediate result. It appears in the diagram in a natural way, as the composition of the intensive quantity of the dollars charged per hour and the extensive quantity of the number of hours. The amount of change is shown in Fig. 3.9 as a difference between the amount paid and the amount charged. Note that there are two expressions for this, one obtained from the expression for the amount charged (4* 0.65) and the other using the numerical amount, 2.60.

Shalin and Bee (1985a) analyzed a class of two-step word problems that are solved using addition, subtraction, multiplication, and division. If consideration is limited to problems with three given quantities, two operations, and one unknown, the problems have diagrams consisting of two triads with one of the quantities in both triads. These problems are of three kinds, differing in the pattern of relations among the quantities. One kind is illustrated by Fig. 3.9, a binary hierarchy, in which the intermediate result is the composed quantity in one triad and one of the components in the other.

Another kind of pattern is illustrated in Fig. 3.10, for the problem "Paul works in a doughnut shop. He made 14 boxes of glazed and

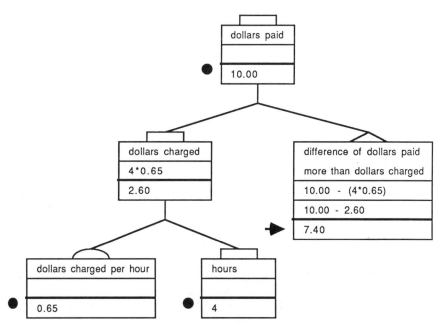

FIG. 3.9. Semantic network for a two-step problem.

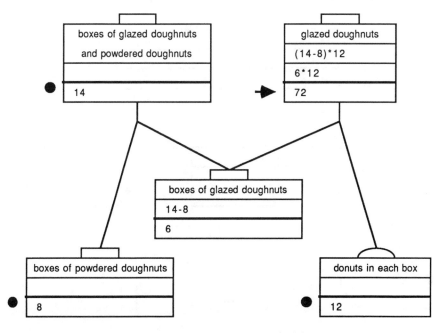

FIG. 3.10. A two-step problem with a shared-component structure.

powdered doughnuts; 8 of these boxes were powdered doughnuts. If he put 12 doughnuts in each box, how many glazed doughnutss did he make?" In this problem the inferred quantity, the boxes of glazed doughnuts, is a component in both of the triads. Its value is obtained by subtracting the number of boxes of powdered doughnuts from the total number of boxes, and then it is used along with the number of doughnuts per box to obtain the number of glazed doughnuts by multiplication.

The third kind of pattern is illustrated in Fig. 3.11, for the problem: "Dr. Wizard has discovered a group of monsters living in a dark cave in South America. He has counted 7 monsters, and there are 8 fingers on each monster. If there are 4 fingers on each monster hand, how many monster hands did he find?" Here the inferred quantity, the total number of fingers, is the combined quantity in both of the triad structures. It is obtained by multiplying the number of monsters by the number of fingers on each monster, and then it is used along with the fingers per hand to obtain the number of hands by division.

A question arises whether the distinctions made by the system are relevant in the problem-solving process. The distinction among the problem structures proposed by Shalin and Bee (1985a)—binary hierarchies, shared components, and shared combinations—is not the standard way of considering differences between problem patterns. In an empirical study, Shalin and Bee (1985b) gave word problems with the patterns they had

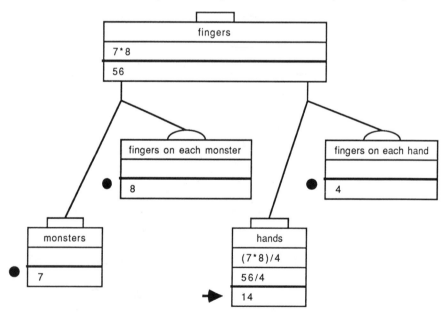

FIG. 3.11.   A problem with a shared-combination structure.

distinguished to third- fourth- and fifth-grade students. They found that student success rate differed for each pattern, although the operations used for the solving the problems were the same. This finding supports the conclusion that the structures of problems, as represented in the diagrams of Shalin's system, are relevant to the process of understanding the problems. It thus encourages exploration of the system as an instructional resource. Shalin and Bee have instructed a few middle-school students informally in use of the semantic-network system to represent information in word problems, and the results have been encouraging. A more systematic, although still preliminary, study was conducted with four college-age students recruited from a remedial mathematics class. These students went through a series of instructional sessions in which they learned to construct semantic networks for one-step problems, involving five different patterns of composition: (a) two extensive quantities composed additively to form another extensive quantity; (b) two extensive quantities related by their additive difference; (c) an extensive and an intensive quantity combined multiplicatively to form another extensive quantity; (d) two extensive quantities related by a multiplicative factor; and (e) two intensive quantities composed multiplicatively to form another intensive quantity. The instruction was quite detailed and required about 15 hour-long sessions.

Protocols were recorded as the students worked on problems during the instruction, and comments in the protocols indicated that the representations helped the students to understand the concepts of quantitative types and relations. The following comment illustrates this:

> The hat, though, tells me that I'm, well, you know, like when you think of math, you think of addition and subtraction together. And multiplication and division together. And if I look at this hat, I know that that's a difference box and difference is subtraction. So, depending on whether, you, though I know I'm going addition or subtraction right there. And then wherever the unknown is, I work toward that. If I'm working up I add, if it's down at the bottom, I'm working down, I subtract.

Shalin obtained additional data by testing students' success at solving word problems without the computational system. Students were given pre- and post-tests, which included the type of word problems they had studied and two-step problems, which they had not. The tests also included problems in both one step and two that indicated the values of quantities with letters rather than numerals, for example, "Barbara and Kenneth have *a* books altogether; *b* of the books are Barbara's; how many of the books belong to Kenneth?"). Problems using letters in this way had not been included in the instruction. No difference was found between pretest and posttest scores obtained for problems presented with numerals, apparently because of a ceiling effect. However, all four students showed improved performance on problems with letters, both in one-step and two-step problems. A reasonable interpretation is that the instruction may have provided students with knowledge for representing problem information more abstractly and so that they had an easier time reasoning about the problems without specific numbers.

Relation of the Representations to Cognitive Models.

Lindvall's system, with dots corresponding to objects, is directly related to the kind of cognitive process hypothesized by Briars and Larkin (1984). Their model simulates construction of a problem representation that uses physical objects or mental tokens corresponding to objects and distingiushes objects by the sets to which they belong.

Shalin's system uses a more abstract representation that resembles the representations of quantities in the schema-based model of Riley et al. (1983), Kintsch and Greeno (1985), and others. Rather than including tokens for individual objects, the system represents each quantity as an object in a diagram whose attributes include the numerical value of the quantity. This corresponds to the process of constructing representations of quantities based on schemata, which is called MAKE-SET in Kintsch and Greeno's version.

The representational system differs from the schema-based model in the way that relations between quantities are represented. They are represented as compositions based on quantitative types rather than on schemata such as Change, Combine, and Compare. The reason for the choice was primarily instructional. When Shalin was designing the instructional representation, the explicit use of schemata seemed cumbersome and likely to distract students from the conceptual level that seemed most likely to help them understand problem information. As a result of Shalin's work, a new hypothesis about the process of understanding has been developed. This hypothesis could lead to a significant extension of cognitive theory about the understanding of quantitative information.

OTHER INSTRUCTIONAL SYSTEMS AND RESEARCH ON UNDERSTANDING

I have discussed in some detail instructional representations based on research about problem understanding. Another domain of tacit knowledge, *strategic knowledge*, also has been used in the development of representational systems for instruction. By strategic knowledge, I mean the knowledge that organizes problem-solving activity, including the processes that set goals and choose plans. Research on problem solving in geometry (Anderson, 1982; Greeno, 1978) and on programming in LISP (Anderson, Farrell, & Sauers, 1984) led to models of the problem-solving process that include setting the subgoals that become the basis of problem-solving activity. Based on these models, Anderson, Boyle, Farrell, & Reiser (1984) leave developed computer-based tutoring systems which include representations of progress towards the solution. A trace of the problem solution is also provided in the form of a graph showing the subgoals and the interfaces made to reach them. The representation is intended to help the student acquire cognitive procedures that are appropriate take problem goals into account as some of the conditions for applying procedures.

Another instructional representation involving strategic knowledge deals with high school algebra and was developed by John Seely Brown and his associates (Greeno et al., 1986). The system ALGEBRALAND displays traces of students' work in solving equations. It provides an opportunity for students to review their work after solving a problem and allows them to reflect on steps that led to dead ends. In addition, students are able to add notations, based on general strategic features of algebraic problem solving like those identified by Bundy (1975), to the problem-solving traces.

Investigators are also making progress exploring students' understanding of the mathematical principles that are relevant to their knowledge of cognitive procedures. Resnick (1982) developed instruction that shows students the detailed correspondence between the steps in written subtraction and the analogous procedure using place-value blocks. Greeno, Resnick and Rowland (cited in Resnick, 1983) developed a plausible model characterizing the understanding that students could acquire from that instruction in terms of schemata of part-whole relations. Although the model was consistent with preliminary observations, more systematic experimentation has shown that the situation is probably more complex. Resnick and Omanson (1984) found that instruction with place-value blocks did not lead all students to perform correctly in paper-and-pencil subtraction, although most students learned to discuss the principles of subtraction more adequately. The relationship between knowledge of principles and the ability to apply cognitive principles correctly requires further investigation.

Another line of research regarding the understanding of procedures involves formal analyses of the implicit understanding of principles that are related to the procedures used to perform cognitive tasks. Gelman and Gallistel (1978) presented evidence that preschool children implicitly understand many principles of number, such as order, one-to-one correspondence, and cardinality, and that this understanding informs and constrains their knowledge of procedures for counting sets of objects. Greeno, Riley, and Gelman (1984) developed a formal analysis of competence that shows how general principles constrain the procedures simulating children's performance. Gelman and Meck (in press) used concepts of competence to interpret the way in which children's performance is influenced by their interpretation of social interactions, and Greeno and Johnson (1984) analyzed the understanding of set-theoretic principles that is involved in understanding and solving arithmetic word problems.

CONCLUSION

The use of cognitive models in the design of instructional resources is in a very early stage. Although existing systems are both interesting and promising, much remains to be understood about the use of these methods and concepts in the development of instruction. Indeed,, they could change instruction in a variety of different ways.

One way would be to provide a new method of teaching something that is already in the curriculum. For example, representations have been developed that could help students succeed in learning to solve word problems in elementary mathematics. Another way these methods and concepts could change instruction is by developing new curricular components. Materials based on these methods could address new instructional objective: helping students learn to understand quantitative information in problem texts. It seems, at least from these examples, that new instructional materials could be developed to use standard tasks in ways that will help reach new curricular objectives.

DISCUSSION

Ron Wenger.   Implicit in the concept of representation is the concept of equality. One of the things that started me thinking about word problems in algebra and elementary functions is the semantics of the drawings—the principles underlying their construction. Can you talk about the correspondence between the drawings and the formal abstract properties of the functions?

Jim Greeno.   I don't have anything very definitive to say about it. The notation fits nicely with notions of compositional equality and those notions fit best in the situations involving subsets and supersets. There is equality there; a superset and its subsets are the same thing. Cases involving changes are more problematic because that fact is not as clear in the physics of the situation. There is an operational result, but that isn't quite the same thing as a compositional equality. In cases involving comparison, there is only numerical equality. The model that the representational system presents is an abstraction from different physical relations into a single notational system. A formal analysis of these problems would include some distinctions that are not included in Shalin's system, but it might not be a good idea to include them all in the instructional system.

Ron Wenger. How can you help students choose from the many ways that two diffferent descriptions contained in a problem can be set equal to yield an equation?

Jim Greeno. The direct answer to your question is that we haven't addressed it yet in our research and design. The main reason is that we have been working with simple problems in which the issue doesn't arise. First of all, for problems like the ones I used as examples that have three numbers given and one unknown, the choice is determined by the problem structure. Two of the given quantities can be used to find the inferred quantity in the pattern, and that quantity can be used with the remaining given quantity to find the answer. For some problems, more than one pattern can be used to represent the quantities and their relations, and the calculations are somewhat different depending on the pattern chosen, but there isn't significant ambiguity about the calculations once a pattern is formed for these simple problems.

Alternative solution sequences are possible for more complex problems, of course. These arise in interesting ways mainly for problems requiring algebra. The problems I used as examples can all be solved using arithmetic operations to infer new numerical values from quantities whose numerical values are given. In problems that require algebra, given numerical values are located in ways that do not allow any direct calculations of values. One of the values has to be called "x" and then has to be propagated through the network as a variable. Somewhere in the network, there is a quantity that has two expressions or one expression and one value, and an equation results from the equality of two items of information. We expect to be able to use the graphics system to teach something about algebraic notation also and to connect it with the semantics of word problems. As we progress with that, we will face some important issues of choosing from the alternative ways of constructing expressions for a problem.

Alan Schoenfeld. It may be worth commenting about the utility of the diagrams that Jim has been discussing. One of the things that has been problematic for math educators and that comes through clearly in the testing literature, such as the National Assessment data, is the difficulty that students have with word problems. Students do fine as long as problems call for only one operation, but as soon as you get to a two-operation word problem, then the kids become confused about which operations to use. Shalin's program serves as more than just a teaching device or a device for structuring the kid's experience. It has the potential to provide the kid with a language or means of describing the reasons for selecting particular numbers and combining them in particular ways. The student has numbers in the boxes; they are connected in a certain way; and the connections correspond to plus, minus, times, or divide. This provides an explicit way for the kids to choose processes to combine the numbers, and there is something concrete for a student and a teacher to have a dialogue about. Once an operation has been selected, there can be a discussion of the rationale for choosing a particular operation. Getting that issue out in the open is bound to be helpful. As I went through my solution of one of the problems Jim used, when I first read it I thought it was a complicated subtraction problem because of the way that it was set up. It seems to me that teachers and students could have useful discussions of the process at this level—"Why choose multiplication?" An important contribution of this kind of schematic device is that it provides one mechanism for students and teachers to discuss the relevance of the appropriate processes.

Jerry Kulm.   I hate to say that this is not new, but I spent some time in Germany where they had a textbook series built on this kind of structure. They called it simplex structure, and they started with kids that are at a very young age.

Jim Greeno.   That's very interesting. I'd like to know more about that.

Jere Confrey.   I feel uncomfortable about the optimistic conjecture that we are doing something significant if we take problems that we give in tests and teach people how to solve them. Sure enough, if you teach kids how to solve textbook word problems, they will do significantly better on those problems, but I doubt that that will help them to solve other kinds of problems.

Jim Greeno.   I agree with your point, as I understand it. The main educational reason for using word problems probably doesn't involve generative mathematical problem solving, but word problems can still have a useful educational purpose. I think that word problems provide a task in which students can learn to understand language that describes quantities and in which the formal symbols of mathematics have more concrete interpretations. These are nontrivial things to learn about, but they are not very generative or productive, in Wertheimer's sense. We need a different set of tasks and experiences for students that will strengthen their ability to think mathematically, to understand the implications of an idea, to explore alternative conjectures. We need tasks that will strengthen other generative capabilities as well.

Ruth Pitt.   How does this kind of representation affect the error rate in the old classical problems like "Write an equation for: 'There are six students for each professor'?"

Jim Greeno.   I don't know. I hope it would be helpful, but tests haven't been conducted. Those ratio problems can be represented with these relational structures and it would be possible to design instruction to teach students how to construct diagrams for problems of that kind. Of course, the success of such instruction would depend heavily on the instruction given as well as the representational system.

Steve Brown.   Have you asked what kind of quantitative structures and what kind of representations people use in the absence of instruction?

Jim Greeno.   We have some data that support characterizations of what people do on their own. [nb. Data are presented by Riley, Greeno, & Heller, 1983; and Shalin & Bee, 1985b.] The evidence is indirect, of course, because as with any understanding process, people are not aware of the representations they are constructing. We are limited, therefore, to indirect evidence.

Steve Brown.   Do you have a systematic method for getting at such evidence?

Jim Greeno.   The data that we have are for youngsters of different ages solving problems with different structures. The model specifies either a more complex process for forming the representation or the need for more knowledge. These differences in the representation match reasonably well with differences in the difficulty that students have solving the problem.

Steve Brown.   Which problems take longer to solve?

Jim Greeno.   That would be relevant. We only have data for the number of students solving each problem.

Jim Kaput.   At the Educational Technology Center in Cambridge we are working on multiplicative word problems along the lines of the question just asked here, trying to find out which are harder and which are easier and why. Judah Schwartz, of course, has taken seriously the distinction between intensive and extensive quantities. He has used the taxonomy generated in the earlier literature involving change problems, comparisons, and so on. On top of that taxonomy is another based on whether the problem is an intensive quantity times an extensive one, an extensive one divided by an extensive one, and so on. We asked students to make up multiplication problems—one-step multiplication problems, one-step division problems, and so on—and then we classified those student responses at a variety of different grade levels and ability levels to see the kinds of problems they made up. Sure enough, the ones that we would expect to be harder are the ones that didn't appear or appeared very infrequently. Then as a second stage we made up tests where the students had not to calculate an answer but to name the appropriate operation. We wanted to find out whether or not they could do these problems, at least up to the point of calculating the answer. There was almost a perfect match between production and comprehension. In fact, we predicted a very tight match, and that's what we got.

The intermediate outcome of this research is a somewhat different type of software. The notion is to look at other semantic features of the quantities involved. For example, in an example involving 153 books and 17 books per shelf, there is obviously a semantic containment relation there because the books are going onto the shelves. That helps steer you toward an appropriate operation. We are looking at systematic variation of that sort of semantic content in the problem, and at familiarity, imagibility, and so on, with a view to setting up a web for each type of problem. Then students can be put into the web, and if they can't solve the problem at one particular place you make it easier for them by backing up to a simpler problem. What I mean by "simpler" is that the problem has been verified as simpler by this earlier empirical work. The idea is that you just keep backing up until you hit one you can solve. The overall notion is that you're really teaching the "think of a simpler problem" heuristic using the computer to help you out.

REFERENCES

Anderson, J. R. (1982). Acquisition of a cognitive skill. *Psychological Review, 89*, 396–406.

Anderson, J. R., Farrell, R., & Sauers, R. (1984). Learning to program in LISP. *Cognitive Science, 8*, 87–130.

Anderson, J. R., Boyle, C. F., Farrell, R., and Reiser, B. J. (1984). *Cognitive principles in the design of computer tutors.* (Report #ONR-84-1), Pittsburgh: Carnegie-Mellon University, Department of Psychology.

Briars, D. J., and Larkin, J. H. (1984). An integrated model of skill in solving elementary word problems. *Cognition & Instruction, 1,* 245–296.

Brown, J. S., & Burton, R. B. (1980). Diagnostic models for procedural bugs in basic mathematical skills. *Cognitive Science, 4,* 379–426.

Bundy, A. (1975). Analyzing mathematical proofs (or reading between the lines). In P. Winston (Ed.), *Proceedings of the 4th International Joint Conference on Artificial Intelligence,* Georgia, USSR.

Carpenter, T. P., & Moser, J. M. (1982). The development of addition and subtraction problem-solving skills. In T. P. Carpenter, J. M. Moser, & T. A. Romberg (Eds.), *Addition and subtraction: A cognitive perspective* (pp. 9–24). Hillsdale, NJ: Lawrence Erlbaum Associates.

Dellarosa, D. (1985). *Solution: A computer simulation of children's arithmetic word problem solving.* (Technical Report No. 148). Boulder: University of Colorado Institute of Cognitive Science.

Fletcher, C. K. (1984). *Understanding and solving word arithmetic problems: A computer simulation.* Boulder: University of Colorado Institute of Cognitive Science.

Gagné, R. M., & Briggs, L. J. (1979). *Principles of instructional design.* New York: Holt Rinehart & Winston.

Gelman, R., & Gallistel, C. R. (1978). *The child's understanding of number.* Cambridge, MA: Harvard University Press.

Gelman, R., & Meck, E. (in press). The notion of principle: The case of counting. In J. Hiebert (Ed.), *Conceptual and procedural knowledge: The case of mathematics.* Hillsdale, NJ: Lawrence Erlbaum Associates.

Greeno, J. G. (1978). A study of problem solving. In R. Glaser (Ed.), *Advances in instructional psychology.* Vol. 1 (pp. 13–75). Hillsdale, NJ: Lawrence Erlbaum Associates.

Greeno, J. R., Brown, J. S., Foss, C., Shalin, V., Bee, N. V., Lewis, M. W., & Vitolo, T. M. (1986). *Cognitive principles of problem solving and instruction.* Berkeley, CA: University of California, School of Education.

Greeno, J. G., & Johnson, W. (1984). *Competence for solving and understanding problems.* Paper presented at the International Congress of Psychology, Acapulco, Mexico.

Greeno, J. G., Riley, M. S., & Gelman, R. (1984). Conceptual competence and young children's counting. *Cognitive Psychology, 16,* 94–143.

Kaput, J. J. (1985). *Multiplicative word problems and intensive quantities: An integrated software response.* Cambridge, MA: Harvard Graduate School of Education, Educational Technology Center.

Kintsch, W., & Greeno, J. G. (1985). Understanding and solving word arithmetic problems. *Psychological Review, 92,* 109–129.

Larkin, J. H., McDermott, J., Simon, D. P., & Simon, H. A. (1980). Models of competence for solving physics problems. *Cognitive Science, 4,* 317–345.

Nesher, P. (1982). Levels of description in the analysis of addition and subtraction word problems. In T. P. Carpenter, J. M. Moser, & T. A. Romberg (Eds.), *Addition and subtraction: A cognitive perspective* (pp. 25–38). Hillsdale, NJ: Lawrence Erlbaum Associates.

Polanyi, M. (1959). *The study of man.* Chicago: University of Chicago Press.

Resnick, L. B. (1982). Syntax and semantics in learning to subtract. In T. P. Carpenter, J. M. Moser, & T. A. Romberg (Eds.), *Addition and subtraction: A cognitive perspective* (pp. 136–155). Hillsdale, NJ: Lawrence Erlbaum Associates.

Resnick, L. B. (1983). A development theory of numbering understanding. In H. P. Ginsburg (Ed.), *The development of mathematical thinking.* New York: Academic Press.

Resnick, L. B., & Omanson, S. F. (1984). *Learning to understand arithmetic.* Pittsburgh: University of Pittsburgh Learning Research and Development Center.

Riley, M. S., & Greeno, J. G. (1985). *Development of understanding*. LaJolla, CA: University of California, San Diego, Center for Cognitive Science.

Riley, M. S., Greeno, J. G., & Heller, J. I. (1983). Development of children's problem-solving ability in arithmetic. In H. P. Ginsburg (Ed.), *The development of mathematical thinking*. New York: Academic Press.

Schwartz, J. (1976). *Semantic aspects of quantity*. Cambridge, MA: MIT, Division for Study and Research in Education.

Schwartz, J. (1982). The semantic calculator. *Classroom Computer News, 2*, 22–24.

Schwartz, J. (1984). *An empirical determination of children's word problem difficulties: Two studies and a prospectus for further research*. Cambridge, MA: Harvard Graduate School of Education, Educational Technology Center.

Shalin, V., & Bee, N. V. (1985a). *Analysis of the semantic structure of a domain of word problems*. Pittsburgh: University of Pittsburgh, Learning Research and Development Center.

Shalin, V., & Bee, N. V. (1985b). *Structural differences between two-step word problems*. Pittsburgh: University of Pittsburgh, Learning Research and Development Center.

Tamburino, J. L. (1982). *The effects of knowledge-based instruction on the abilities of primary grade children in arithmetic word problem solving*. Unpublished doctoral dissertation, University of Pittsburgh.

Thibadeau, R., Just, M. A., & Carpenter, P. A. (1982). A model of the time course and content of reading. *Cognitive Science, 6*, 157–203.

Thorndike, E. L. (1922). *The psychology of arithmetic*. New York: Macmillan.

van Dijk, T. A., & Kintsch, W. (1983). *Strategies of discourse comprehension*. New York: Academic Press.

Vergnaud, G. (1982). A classification of cognitive tasks and operations of thought involved in addition and subtraction problems. In T. P. Carpenter, J. M. Moser, & T. A. Romberg (Eds.), *Addition and subtraction: A cognitive perspective* (pp. 39–59). Hillsdale, NJ: Lawrence Erlbaum Associates.

Vergnaud, G. (1983). Multiplicative structures. In R. Lesh & M. Landau (Eds.), *Acquisition of mathematics concepts and processes* (pp. 128–175). New York: Academic Press.

Waller, D. M. (1983). *Implementation of a method for teaching a solution of arithmetic story problems using diagrammatic models*. Unpublished master's thesis, University of Pittsburgh.

4 Cognitive Technologies for Mathematics Education

Roy D. Pea
Educational Communication and Technology
New York University

This chapter begins with a sociohistorical perspective on the roles played by cognitive technologies as reorganizers rather than amplifiers of mind. Informed by patterns of the past, perhaps we can better understand the transformational roles of advanced technologies in mathematical thinking and education. Computers are doing far more than making it easier or faster to do what we are already doing. The sociohistorical context may also illuminate promising directions for research and practice on computers in mathematics education and make sense of the drastic reformulations in the aims and methods of mathematics education wrought by computers.

The chapter then proposes an heuristic taxonomy of seven functions whose incorporation into educational technologies may promote mathematical thinking. It distinguishes two types of functions: *purpose* functions, which may affect whether students choose to think mathematically, and *process* functions, which may support the component mental activities of mathematical thinking. My hope is that the functions falling into these two categories will apply to all cognitive technologies, that they will help students to think mathematically, and that they can be used both retroactively to assess existing software and proactively to guide software development efforts. Definitions and examples of software are provided throughout the chapter to illustrate the functions.

The central role that mathematical thinking should play in mathematics education is now receiving more attention, both among educators and in the research community (e.g. Schoenfeld, 1985a; Silver, 1985). As Schoenfeld says, "You understand how to think mathematically when you are resourceful, flexible, and efficient in your ability to deal with new problems in mathematics" (1985a, p. 2). The growing alignment of mathematics learning with mathematical thinking is a significant shift in education.

THEMES OF CHANGE IN THE AIMS AND METHODS OF MATH EDUCATION

There is no question but that information technologies, in particular the computer, have radical implications for our methods and are already changing them. But, perhaps more importantly, we are also coming to see that they are changing our *aims* and thereby what we consider the goals of mathematical understanding and thinking to which our educational processes are directed.

Mathematics educators, represented by such organizations as NCTM, are fundamentally rethinking their aims and means. In particular, mathematics activities are becoming significant in a much wider variety of contexts than ever before. The reason for this expansion is the widespread availability of powerful mathematical tools that simplify numerical and symbolic calculations, graphing and modeling, and many of the mental operations involved in mathematical thinking. For example, many classrooms now have available programmable calculators, computer languages, simulation and modeling languages, spreadsheets, algebraic equation solvers such as TK!SOLVER, symbolic manipulation packages and software for data analysis and graphing. The drudgery of remembering and practicing cumbersome algorithms is now often supplanted by activities quite different in nature: selecting appropriate computer programs and data entry.

Why have these revolutionary changes occurred? How can we use them as a guide in the design, testing, and use of the new technologies, so that we can enhance both the processes of mathematics education and our understanding of how it occurs? In other words, what are the beacons that will help light the way as we consider the role of cognitive technologies in mathematics education?

AN HISTORICAL APPROACH TO MENTAL ROLES FOR COGNITIVE TECHNOLOGIES

An historical approach will help us consider how the powers of information technologies can best serve mathematics education and research. It will help us look beyond the information age to understand the transformational roles of cognitive technologies and to illuminate their potential as tools of mentation. Long before computers appeared, technical instruments such as written language expanded human intelligence to a remarkable extent. I take as axiomatic that intelligence is not a quality of the mind alone, but a product of the relation between mental structures and the tools of the intellect provided by the culture (Bruner, 1966; Cole & Griffin, 1980; Luria, 1976, 1979; Olson, 1976, 1985; Olson & Bruner, 1974; Pea, 1985b; Vygotsky, 1962, 1978). Let us call these tools cognitive technologies.

A cognitive technology is any medium that helps transcend the limitations of the mind (e.g., attention to goals, short-term memory span) in thinking, learning, and problem-solving activities. Cognitive technologies have had remarkable consequences on the varieties of intelligence, the functions of human thinking, and past intellectual achievements (e.g., Cassirer, 1944; Goodman, 1976). They include all symbol systems, including writing systems, logics, mathematical notation systems, models, theories, film and other pictorial media, and now symbolic computer languages. The technologies that have received perhaps the most attention as cognitive tools are *written language* (Goody, 1977; Greenfield, 1972; Olson, 1977; Ong, 1982; Scribner & Cole, 1981), and *systems of mathematical notation*, such as algebra or calculus (Cassirer, 1910, 1957; Kaput, 1985, in press; Kline, 1972) and number symbols (Menninger, 1969).

Contrast for a moment what it meant to learn math with a chalk and board, where one erased after each problem, with what it meant to use paper and pencil, where one could save and inspect one's work. This example reminds us that under the broad rubric of the "cognitive technologies" for mathematics, we must include entities as diverse as the chalk and board, the pencil and paper, the computer and screen, and the symbol systems within which mathematical discoveries have been made and that have led to the creation of new symbol systems. *Each has transformed how mathematics can be done and how mathematics education can be accomplished.* It would be interesting to explore, if space allowed, the particular ways in which mathematics and mathematics education changed with the introduction of each medium.

A common feature of all these cognitive technologies is that they make *external* the intermediate products of thinking (e.g., outputs of component steps in solving a complex algebraic equation), which can then be analyzed, reflected upon, and discussed. Transient and private thought processes subject to the distortions and limitations of attention and memory are "captured" and embodied in a communicable medium that persists, providing material records that can become objects of analysis in their own right—conceptual building blocks rather than shifting sands. Vygotsky (1978) heralded these tools as the "extracortical organizers of thought," because they help organize thinking outside the physical confines of the brain.

We are now seeing, in ways described throughout the chapter, how computers are an especially potent type of cognitive technology for learning to think mathematically: They can operate not only with numbers, but also with symbols—the fundamental currency of human thought. Computers are universal machines for storing and dynamically manipulating symbols. Capable of real-time programmable interactions with human users, computers may provide the most extraordinary cognitive technologies thus devised. But what can we learn from the history of noncomputer-based cognitive technologies that will inform our current inquiries?

Cognitive technologies, such as written languages, are commonly thought of as *cultural amplifiers* of the intellect, to use Jerome Bruner's (1966, p. xii) phrase. They are viewed as cultural means for empowering human cognitive capacities. Greenfield and Bruner (1969) observed that cultures with technologies such as written language and mathematical formalisms will "push cognitive growth better, earlier, and longer than others" (p. 654). We find similarly upbeat predictions embodied in a widespread belief that computer technologies will inevitably and profoundly amplify human mental powers (Pea & Kurland, 1984).

This amplifier metaphor for cognitive technologies has led to many research programs, particularly on the cognitive consequences of literacy and schooling (e.g. on formal logical reasoning) in the several decades since Bruner and his colleagues published *Studies in Cognitive Growth* (e.g., Greenfield, 1972; Olson, 1976; Scribner & Cole, 1981). The metaphor persists in the contemporary work on electronic technologies by John Seeley Brown of Xerox PARC, who, in a recent paper, described his prototype software systems for writing and doing mathematics as "idea amplifiers" (Brown, 1984a). For example, AlgebraLand, created by Brown and his colleagues (Brown, 1984b), is a software program in which students are freed from hand calculations associated with executing different algebraic operations and allowed to focus on high level problem-solving strategies they select for the computer to perform. AlgebraLand is said to enable students "to explore the problem space faster," as they learn equation solving skills. Although *quantitative* metrics, such as the efficiency and speed of learning, may truly describe changes that occur in problem solving with electronic tools, more profound changes—as I will later describe for the AlgebraLand example—may be missed if we confine ourselves to the amplification perspective.

There is a different tradition that may be characterized as the *cultural-historical* study of cognitive technologies. This perspective is most familiar to psychologists and educators today in the influential work of Vygotsky (1978). Vygotsky offered an account of the development of higher mental functions, such as planning and numerical reasoning, as being baed on the "internalization" of self-regulatory activities that first take place in the social interaction between children and adults. The historical roots of Vygotsky's orientation provide an illuminating framework for the roles of computer technologies in mathematical thinking and learning. Influenced by the writings of Vico, Spinoza, and Hegel, Marx and Engels developed a novel and powerful theory of society now described as *historical,* or *dialectical, materialism.* According to this theory, human nature is not a product of environmental forces, but is of our own making as a society and is continually in the process of "becoming." Humankind is reshaped through a dialectic, or "conversation," of reciprocal influences: *Our productive activities change the world, thereby changing the ways in which the world can change us.* By shaping nature and how our interactions with it are mediated, we change ourselves. As the biologist Stephen Jay Gould observes (1980), such "cultural evolution," in contrast to Darwinian biological evolution, is defined by the transmission of skills, knowledge, and learned behavior across generations. It is one of the ways that we as a species have transcended nature.

Seen from this cultural-historical perspective, *labor* is the factor mediating the relationship of human beings to nature. By creating and using physical instruments (such as machinery) that make our interaction with nature less and less direct, we reshape our own, human nature. The change is fundamental: Using different instruments of work (e.g. a plow rather than the hand) changes the functional organization, or system characteristics, of the human relationship to work. Not only is the work finished more quickly, but the actions necessary to accomplish the required task have changed.

In an attempt to integrate accounts of individual and cultural changes, the Soviet theorists L. S. Vygotsky (e.g., 1962, 1978) and A. R. Luria (1976, 1979) generalized the historical materialism that Marx and Engels developed for physical instruments. They applied it to an historical analysis of symbolic tools, such as written language, that serve as instruments for redefining culture and human nature. What Vygotsky recognized was that "mental processes always involve signs, just as action on the environment always involves physical instruments (if only a human hand)" (Scribner & Cole, 1981, p. 8). A similar instrumental and dialectical perspective is reflected in recent studies of the "child as a cultural invention" (Kessel & Siegel, 1983; Kessen, 1979; White, 1983). Take, for instance, Wartofsky's (1983) description of the shift in perspective:

> Children are, or become, what they are taken to be by others, and what they come to take themselves to be, in the course of their social communication and interactions with others. In this sense, I take "child" to be a social and

historical kind, rather than a natural kind, and therefore also a constructed
kind rather than one given, so to speak, by nature in some fixed or essential
form. (p. 190)

Applied to mathematics education, this sociohistorical perspective
highlights not the *constancy* of the mathematical understandings of which
children are capable at particular ages, but how what we take for granted
as limits are redefined by the child's use of new cognitive technologies for
learning and doing mathematics. Similarly, Cole and Griffin (1980) noted
how symbolic technologies qualitatively change the structure of the
functional system for such mental activities as problem solving or memory.

The term "amplify" has other implications. It means to make more
powerful, and to amplify in the scientific sense "refers specifically to the
intensification of a signal (acoustic, electronic), *which does not undergo
change in its basic structure*" (Cole & Griffin, 1980, p. 349). Thus, the
amplifier metaphor for the roles of technologies in mathematical thinking
leads one to unidimensional, quantitative theorizing about the effects of
cognitive technologies. A pencil seems to amplify the power of a sixth
grader's memory for a long list of words when only the *outcome* of the list
length is considered. But it would be distortive to go on to say that the
mental process of remembering that leads to the outcome is amplified by
the pencil. The pencil does not amplify a fixed mental capacity called
memory; it restructures the functional system of remembering and
thereby leads to a more powerful outcome (at least in terms of the number
of items memorized). Similar preoccupations with amplification led researchers to make quantitative comparisons of enhancements in the
learning of basic math facts that are brought about by software and print
media, rather than to consider the fundamental changes in arithmetical
thinking that accompany the usage of programmable calculator functions
(Conference Board of the Mathematical Sciences, 1983; Fey, 1984; National Science Board, 1983).

Olson (1976) makes similar arguments about the capacity of written
language to restructure thinking processes. For example, written language facilitates the logical analysis of arguments for consistency–contra-
diction because print provides a means of storing and communicating
cultural knowledge. It transcends the memory limitations of oral language. What this means is that technologies do not simply either *amplify*,
like a radio amplifier, the mental powers of the learner or speed up and
make the process of reaching previously chosen educational goals more
efficient. The standard image of the cognitive effects of computer use is
one-directional: that of the child seated at a computer terminal and
undergoing certain changes of mind as a direct function of interaction
with the machine. The relatively small number of variables to measure

makes this image seductive for the researcher. But since the technologies change the system of thinking activities in which the technologies play a role, their effects are much more complex and often indirect. Like print, they transcend the memory limitations of oral language. Complicating matters even more is that the *specific* restructurings of cognitive technologies are seldom predictable; they have emergent properties that are discovered only through experimentation.

I espouse a quite different theory about the cognitive effects of computers than that just described. My theory is consistent with questions based in a two-directional image that other mathematics educators and researchers (e.g., Kaput, 1985, in press) are posing, such as: What are the *new* things you can do with technologies that you could not do before or that weren't *practical* to do? Once you begin to use the technology, what totally new things do you realize might be *possible* to do? By "two-directional image," I mean that not only do computers affect people, but people affect computers. This is true in two senses. In one sense, we all affect computers and the learning opportunities they afford students in education by how we *interpret* them and by what we define as appropriate practices with them; as these interpretations change over time, we change the effects the computers can have by changing what we do with them. (Consider how we began in schools, with drill and practice and computer literacy activities, and now emphasize the uses of computers as tools, such as word processors, spreadsheets, database management systems.) In another sense, we affect computers when we study their use, reflect on what we see happening, and then act to change it in ways we prefer or see as necessary to get the effects we want. Such software engineering is fundamentally a dialectical process between humans and machines. We define the educational goals (either tacitly or explicitly) and then create the learning activities that work toward these goals. We then try to create the appropriate software. We experiment and test, experiment and test, until we are satisfied . . . which we tend never to be. Experimentation is a spiral process toward the unknown. Through experimentation, new goals and new ideas for learning activities emerge. And so on it goes—we create our own history by remaking the tools with which we learn and think, and we simultaneously change our goals for their use.

COGNITIVE TECHNOLOGIES IN MATHEMATICS EDUCATION

How does the idea of cognitive technologies relate to mathematics education? A few historical notes prepare the stage. We may recall Ernst Mach's (1893/1960) statement, in his seminal work on the science of mechanics earlier this century, that the purpose of mathematics should be to save mental effort. Thus arithmetic procedures allow one to bypass counting procedures, and algebra substitutes "relations for values, symbolizes and definitively fixes all numerical operations that follow the same rule" (p. 583). When numerical operations are symbolized by mechanical operations with symbols, he notes, "our brain energy is spared for more important tasks" (p. 584), such as discovery or planning. Although overly neural in his explanation, his point about freeing up mental capacity by making some of the functions of problem solving automatic is a central theme in cognitive science today.

Whitehead (1948) made a similar point: "By relieving the brain of all unnecessary work, a good notation sets it free to concentrate on more advanced problems, and in effect increases the mental power" (p. 39). He noted that a Greek mathematician would be astonished to learn that today a large proportion of the population can perform the division operation on even extremely large numbers (Menninger, 1969). He would be more astonished still to learn that with calculators, knowledge of long division algorithms is now altogether unnecessary. Further arguments about the transformational roles of symbolic notational systems in mathematical thinking are offered by Cajori (1929a, 1929b), Grabiner (1974), and, particularly, Kaput (in press).

Although long on insight, Mach and Whitehead lacked a cognitive psychology that explicated the *processes* through which new technologies could facilitate and reorganize mathematical thinking. What aspects of mathematical thinking can new cognitive technologies free up, catalyze, or uncover? The remainder of this chapter is devoted to exploring this central question.

A historical approach is critical because it enables us to see how looking only at the contemporary situation limits our thinking about what it *means* to think mathematically and to be mathematically educated (cf. Resnick & Resnick, 1977, on comparable historical redefinitions of "literacy" in American education). These questions become all the more significant when we realize that our cognitive and educational research conclusions to date on what student of a particular age or Piagetian developmental level can do in mathematics are restricted to the *static* medium of mathematical thinking with paper and pencil.[1] The dynamic and interactive media provided by computer software make gaining an intuitive understanding (traditionally the province of the professional mathematician) of the interrelationships among graphic, equational, and pictorial representations more accessible to the software user. Doors to mathematical thinking are opened, and more people may wander in.

---

[1]This argument is developed more fully with respect to cognitive development in general with new technologies in Pea (1985a).

Thus, the basic findings of mathematical education will need to be rewritten, so that they do not contain our imagination of what students might do, thereby hindering the development of new cognitive technologies for mathematics education.

TRANSCENDENT FUNCTIONS FOR COGNITIVE TECHNOLOGIES IN MATHEMATICS EDUCATION

Rationale

What strategy shall we choose for thinking about and selecting among cognitive technologies in mathematics education? I argue for the need to move beyond the familiar cookbooks of 1,001 things, in near random order, that one *can* do with a computer. Such lists are usually so vast as to be unusable in guiding the current choice and the future developments of mathematics educational technologies. Instead, we should seek out high leverage aspects of information technologies that promote the development of mathematical thinking skills. I thus propose a list of "transcendent functions" for cognitive technologies in mathematics education.

What is the status of such a list of functions? Incorporating them into a piece of software would certainly not be sufficient to promote mathematical thinking. The strategy is more *probabilistic*—other things being equal, more students are likely to think mathematically more frequently when technologies incorporate these functions. Some few students will become prodigious mathematical thinkers, whatever obstacles must be overcome in the mathematics education they face.[2] Others will not thrive without a richer environment for fostering mathematical thinking. This taxonomy is designed to serve as a heuristic, or guide. Assessments of whether it is useful will emerge from empirical research programs, not from intuitive conjecture. Indeed, until tighter connections can be drawn between theory and practice,[3] the list can only build on what we know from research in the cognitive sciences; it should not be limited by that research.

---

[2]It is more commonly true that prodigious mathematical thinkers have had a remarkable coalescence of supportive environmental conditions for their learning activities, e.g., suitable models, rich resource environment of learning materials, community of peers, and private tutoring (e.g., as described by Feldman, 1980).

[3]This situation is the rule in theory-practice relations in education (Champagne & Chaiklin, 1985; Suppes, 1978). For this reason, I have recently proposed (Pea, 1985b) the need for an activist research paradigm in educational technology, with the goal of simultaneously creating and studying changes in the processes and outcomes of human learning with new cognitive and educational tools.

Finally, why should we focus on transcendent functions? There are two major reasons. We would like to know what functions can be common to all mathematical cognitive technologies, so that each technology need not be created from the ground up, mathematical domain by mathematical domain. We would like the functions to be transcendent in the sense that they apply not only to arithmetic, or algebra, or calculus, but potentially across a wide array, if not all, of the disciplines of mathematical education, past, present, and future. The transcendent functions of mathematical cognitive technologies should thus survive changes in the K–12 math curricula, *since they exploit general features of what it means to think mathematically*—features that are at the core of the psychology of mathematics cognition and learning. These functions should be central regardless of the career emphasis of the students and regardless of their academic future. Lessons learned about these functions from research and practice should allow productive generalizations.

The transcendent functions to be highlighted are those presumed to have great impact on mathematical thinking. They neither begin nor end with the computer but arise in the course of teaching, as part of human interaction. Educational technologies thus only have a role within the contexts of human action and purpose. Nonetheless, interactive media may offer extensions of these critical functions. Let us consider what these extensions are and how they make the nature or variety of mathematical experience qualitatively different and more likely to precipitate mathematics learning and development.

These functions are by no means independent, nor is it possible to make them so. They define central tendencies with fuzzy boundaries, like concepts in general (Rosch & Mervis, 1975). They are also not presented in order of relative importance. I will illustrate by examples how many outstanding, recently developed mathematical educational technologies incorporate many of the functions. But very few of these programs reflect *all* of the functions. And only rare examples in classical computer-assisted instruction, where electronic versions of drill and practice activities have predominated, incorporate any of the functions.

One could approach the question of technologies for math education in quite different ways than the one proposed. One might imagine approaches that assume the dominant role for technology to be amplifier: to give students *more* practice, more *quickly,* in applying algorithms that can be carried out *faster* by computers than otherwise. One could discuss the best ways of using computers for teacher record-keeping, preparing problems for tests, or grading tests. In none of these approaches, however, can computers be considered *cognitive* technologies.

A different perspective on the roles of computer technologies in mathematics education is taken by Kelman et al. (1983) in their book, *Computers in Teaching Mathematics*. They describe various ways software can help create an effective environment for student problem solving in mathematics. Their comprehensive book is organized according to traditional software categories and curriculum objectives: computer-assisted instruction, problem solving, computer graphics, applied mathematics, computer science, programming and programming languages. The spirit of their recommendations is in harmony with the sketch I propose in this chapter, although their orientation is predominantly curricular rather than cognitive. My stress on transcendent functions is thus a complementary approach, taking as a starting point the root or foundational psychological processes embodied in software that engages mathematical thinking.

In my choice of software illustrations I have leaned heavily toward cases that manifest most clearly the specific loci supporting the seven Purpose or Process functions. Although programming languages, spreadsheets, simulation modeling languages such as MicroDynamo (Addison-Wesley), and symbolic calculators such as muMath (MicroSoft) and TK!Solver (Software Arts) can be central to thinking mathematically in an information age (e.g., Elgarten, Posamentier, and Moresh, 1983), I have seldom chosen them as examples. Although I take for granted the utility and power of these types of tools in the hands of a person committed to problem solving, their usefulness stems in part from the extent to which they incorporate the purpose and process functions. For example, Logo graphics programming provides the different mathematical representations of procedural text instructions and the graphics drawing it creates (Process Function 3); and simulation modeling languages and spreadsheets are excellent environments for mathematical exploration (Process Function 2), since hypothesis-testing and model development and refinement are central uses of these interactive software tools. But other environments in which these tools are used—for example, drill and practice on programming language syntax or abstract exercises to write programs to create fibonnaci number series need not offer much encouragement for mathematical thinking. In other words, the intrinsic value of such tools in helping students think mathematically is not a given. The stress on Functions remains central.

A GUIDING DICHOTOMY: PURPOSE AND PROCESS FUNCTIONS FOR COGNITIVE TECHNOLOGIES

How can technology support and promote thinking mathematically? In broad strokes, what appear to be the richest loci of potential cognitive and motivational support of technologies for math education?

We can think of two sides to the educational practices of mathematics learning and ask how software can help. The first side is the personal side—will students choose to commit themselves to learning to think mathematically? Mathematics educators have to some extent neglected the concepts of motivation and purpose (e.g., McLeod, 1985); that neglect may help explain girls' and minorities' documented lack of interest in mathematics. What students learn also depends on the *cognitive support* given them as they learn the many problem-solving skills involved in thinking mathematically.

My perspective on the functions necessary for cognitive technologies thus has two vantage points. First, students are purposive, goal-directed learners, who have the will (on any given occasion or over time) to learn to think mathematically or not. Then once they have embarked on mathematical thinking, they may be aided by technologies in mathematical thinking. For simplicity of exposition, we thus divide function types between: (a) those which promote PURPOSE—engaging students to think mathematically; and (b) those which promote PROCESS—aiding them once they do so.

Purpose Functions in Cognitive Technologies

What lies at the heart of cognitive technologies that help make mathematical thinking purposeful and help commit the learner to the pursuit of understanding? Cognitive technologies that accomplish these goals are based on a participatory link between self and knowledge rather than an arbitrary one. This organic relationship was central to John Dewey's pedagogical writings and integral to Piaget's constructivism: We must build on the child's interests, desires and concerns, and more generally, on the child's world view. But what exactly does this mean?

The key idea behind purpose functions is that they promote the formation of promathematics belief systems in students and thus ensure that students become mathematical thinkers who participate in and own what is learned. Students benefiting from purpose functions are no longer mere storage bins for or executors of "someone else's math." The implication is that technologies for mathematics education should be tools for promoting the student's self-perception as mathematical "agent," as subject or creator of mathematics (Papert, 1972, 1980). For example, Schoenfeld (1985a, 1985b) argues that the belief systems an individual holds can dramatically influence the very *possibilities* of mathematical education:

> Students abstract a "mathematical world view" both from their experiences with mathematical objects in the real world and from their classroom experiences with mathematics. . . . These perspectives affect the ways that students behave when confronted with a mathematical problem, both

influencing what they perceive to be important in the problem and what sets of ideas, or cognitive resources, they use (Schoenfeld, 1985, p. 157).

Although Schoenfeld's focus is broader than the point here, the student's mathematical world view includes the *self*: What am I in relation to this mathematical behavior I am producing? If students do not view themselves as mathematical thinkers, but only as recipients of the "inert" mathematical knowledge that others possess (Whitehead, 1929), then math education *for* thinking is going to be problematical—because the agent is missing.

In the prototypical educational setting, we often erroneously presuppose that we have engaged the student's learning commitment. But the student rarely sees significance in the learning; someone else has made all the decisions about scope and sequence, about the lesson for the day. The learning is meant to deal not with the student's problem or a problematic situation the teacher has helped highlight, but with someone else's. And the knowledge used to solve the problem is someone else's as well, something that someone else might have found useful at some other time. Even that past utility is seldom conveyed: students are almost never told how measurement activities were essential to building projects or making clothes, or how numeration systems were necessary for trade (McLellan & Dewey, 1895).

According to Dewey's (1933, 1938) scheme for the logic of inquiry, the prototypical system of delivering mathematical facts leaves out the necessary *first step* in problem solving: the identification of the problem, the tension that arises between what the student already knows and what he or she needs to know that drives subsequent problem-solving processes. It is interesting that Pólya (1957) also omits this first step; in other respects his phases of problem solving correspond to Dewey's seminal treatment: problem definition, plan creation, plan execution, plan evaluation, and reflection for generalization of what can be learned from this episode for the future (cf., Noddings, 1985). Perhaps the expert mathematician takes this first step for granted: For who could not notice mathematical problems? The world is full of them! But for the child meeting the formal systems that mathematics offers and the historically accrued problem-solving contexts for which mathematics has been found useful, the first step is a giant one, requiring support.

Purpose functions that help the student become a thinking subject can be incorporated into mathematically oriented educational technologies in many ways. Here, we go beyond Dewey to suggest other component features of mathematical agency:

1. *Ownership*. Agency is more likely when the student has primary ownership of the problem for which the knowledge is needed (or second-

ary ownership, i.e., identification with the actor in the problem setting, in an "as if it were me" simulation). A central pedagogical concern is to find ways to help people "own" their own thoughts and the problems through which they will learn. Kaput (1985) and Papert (1980) have provided suggestive examples from software mathematics discovery environments where the "epistemological context" is redefined: Authority for what is known must rest on proof by either the student or the teacher; it must not rest exclusively with the teacher and the text. Students can offer new problems to be solved, and they can also create new knowledge.

2. *Self-worth.* It is hard for students to be mathematical agents if they view opportunities for thinking as occasions for failure and diminished self-worth. Student performance depends partly on self-concept and self-evaluation (Harter, 1985). Research on the motivation to achieve by Dweck and colleagues (e.g., Dweck & Elliot, 1983) indicates that students tend to hold one of two dominant views of intelligence, and that the one held by each particular student helps determine his or her goals. On one hand, if the child views intelligence as an entity, a given quantity of something that one either has or has not, then the learning events arranged at school become opportunities for success or occasions for failure; if the child looks bad, his or her self-concept is negatively affected. On the other hand, if the child views intelligence as "incremental," then these same learning events are viewed as opportunities for acquiring new understanding. Although little is known about the ontogenesis of the detrimental entity view, it is apparent that this belief can hinder the possibility of mathematical agency and that software or thinking practices that foster an incremental world view should be sought.

3. *Knowledge for action.* A third condition for promoting mathematical agency is either that the mathematical knowledge and skills to be acquired have an impact on students' own lives or future careers or that knowledge actually facilitates their solution of real-world problems. New knowledge, whether problem-solving skills or new mathematical ideas, should EMPOWER children to understand or do something better than they could prior to its acquisition. That this condition is important is clear from research on the transferability of instructed thinking skills such as memory strategies (e.g., Campione & Brown, 1978). This research indicates easier transfer of the new skills to other problem settings if one simply explains the benefits of the skill to be learned, that is, that more material will be remembered if one learns this strategy.

Technologies for mathematical thinking that incorporate these Purpose functions should make clear the impact of the new knowledge on the students' lives.

To summarize: In characterizing the general category of Purpose functions for cognitive technologies, I have focused on the importance of linking the child-as-agent with the knowledge to be acquired instead of on the alleged motivational value (e.g., Lepper, 1985; Malone, 1981) of mathematical educational technologies. I have done so because it is inappropriate to think about technologies as artifacts that mechanistically induce motivation. That perspective has led to the extrinsic motivation characterizing most current learning-game software: bells and whistles are added that serve no function in the student's mathematical thinking. Furthermore, these extrinsic motivational features are not proagentive in the sense described earlier. Incorporating the purpose functions I have described into educational technologies could help strengthen *intrinsic* motivation. This can be done by building educational technologies based in specific types of functional and social environments.

Functional Environments That Promote Mathematical Thinking

These are environments that help motivate students to think mathematically by providing mathematics activities whose purposes go beyond "learning math." Whole problems, in which the mathematics to be learned is essential for dealing with the problems, are the focus. The mathematics becomes *functional,* since the technologies prompt the development of mathematical thinking as a means of solving a problem rather than as an end in itself. Systems that provide a functional environment help students interpret the world mathematically in a problem-solving context. Just as in real-life problem solving, associated curricula are not disembodied from purpose (Lesh, 1981). In other words, students see that the mathematics used has a point and can join in the learning activities that pursue the point.

An example is provided in a three-stage approach to algebra education using new technologies outlined in the recent *Computing and Education Report* (Fey, 1984, p. 24). In Stage 1, students begin with "problem situations for which algebra is useful." These types of problem situations—such as science problems of projectile motion and nonlinear profit or cost functions—offer "the best possible motivations." In Stage 2, they learn how to solve such problems using guess-and-test successive approximations—by hand, by graph, and by computer—as well as by means of formal computer tools such as TK!Solver and muMath. In Stage 3, they learn more formal techniques for solving quadratics, such as factoring formulas and the number and types of possible roots. Through such a sequence, students begin by seeing several applications immediately, not by learning techniques whose applications they will see only later. Similar sequences developed from mathematically complex musical or artistic creations are also possible.

Although such functional environments for learning mathematics can be created without computers, computers widen our options. Software may provide innovative, adventurelike problem-solving programs for which mathematical thinking is required of the players if they are to succeed. The five programs in Tom Snyder's (1982) *Search Series* (Mc-Graw-Hill), for example, encourage group problem solving. In Energy Search, students manage an energy factory, collaboratively making interdependent decisions to seek out new energy sources. Geography Search sets students off on a New World search for the Lost City of Gold; climate, stars, suns, water depth, and wind direction, availability of provisions, location of pirates, and other considerations must figure in their navigation plans and progress.

In Bank Street's multimedia "Voyage of the Mimi" Project in Science and Math Education (Char, Hawkins, Wootten, Sheingold, & Roberts, 1983) video, software, and print media weave a narrative tale of young scientists and their student assistants engaged in whale research. Science problems and uses for mathematics and computers emerge and are tackled cooperatively during the group's adventures. One of the software programs, Rescue Mission (also created by Tom Snyder), simulates navigational instruments—such as radar and a direction finder—used on the Mimi vessel and the realistic problem of how to use navigation to save a whale trapped in a fishing net. To work together effectively during this software game, students need to learn how to plan and keep records of emerging data, work on speed-time-distance problems, reason geometrically, and estimate distances. It is in the context of needing to do these things that mathematics comes to serve a functional role.

Sunburst Corporation has also published numerous programs that highlight simulations of real-life events in which students use mathematics skills as aids to planning and problem solving in real-world situations. For example, Survival Math requires mathematical reasoning to solve real-world problems such as shopping for best buys, trip planning, and building construction, and The Whatsit Corporation requires students to run a business producing a product. While problems such as these can be solved on paper, the interactive, model-building features of the computer programs can motivate mathematical thinking much more effectively.

Social Environments for Mathematical Thinking

Social environments that establish an *interactive social context* for discussing, reflecting upon, and collaborating in the mathematical thinking necessary to solve a problem also motivate mathematical thinking.

Studies of mathematical problem solving, for example, by Noddings (1985), Pettito (1984a, 1984b), and Schoenfeld (1985b) indicate how useful *dialogues* among mathematics problem solvers can be in learning to think mathematically. Small group dialogues prompt disbelief, challenge, and the need for explicit mathematical argumentation; the group can bring more previous experience to bear on the problem than can any individual; and the need for an orderly problem-solving process is highlighted (Noddings, 1985). Cooperative learning research in other disciplines of schooling (e.g., Slavin, 1983; Slavin et al., 1985; Stodolsky, 1984) and the new focus in writing composition instruction that emphasizes thinking-aloud activities (Bereiter & Scardamalia, 1986) also focus on social environments. The computer can serve as a fundamental *mediational* tool for promoting dialogue and collaboration on mathematical problem solving. In mathematical learning, as in writing process activities (Grave & Stuart, 1985; Mehan, Moll, & Riel, 1985), social contexts can open up opportunities for the child to develop a distinctive "voice" and to internalize the critical thinking processes that get played out socially in dialogue.

To date, computers have rarely been used to facilitate this function explicitly. The record-keeping and tool functions of software could, however, effectively support collaborative processes in mathematics, just as they have in multiple text authoring environments (Brown, 1984b). This function is usually exploited only implicitly, as in Logo programming, where students often work together to create a graphics program. In doing so, they argue the comparative merits of strategies for solving the mathematics problems that are involved in the programming (Hawkins, Hamolsky, & Heide, 1983; Webb, 1984). The public nature of the computer screen and the ease of revision further encourage collaboration among students (Hawkins, Sheingold, Gearhart, & Berger, 1982). Self-esteem can also grow in a collaborative context when students view one of their peers as expert. There have been some instances in Logo programming research where students with little previous peer support and low self-esteem have emerged as "experts" (Sheingold, Hawkins, & Char, 1984; Papert, Watt, diSessa, & Weir, 1979).

Mathematics is often a social activity in the world. Explicitly recognizing and encouraging this in mathematics education would not only be educationally beneficial and more realistic, but would also make mathematics more enjoyable—sharable rather than sufferable. Mathematics educators should provide better tools for collaborating in mathematics problem solving and work towards promoting more instructionally relevant peer dialogue around mathematical thinking activities.

An example of a program that does just that is part of the Voyage of the Mimi Project in Science and Math Education at Bank Street (Char, Hawkins, Wootten, Sheingold, & Roberts, 1983), a line of software called the Bank Street Lab, developed in conjunction with TERC. It is composed of various kinds of group activities for conducting experiments involving Probe, a hardware device that plugs into the microcomputer and can measure and graph changes in light, heat, temperature, and sound over time. Students work together taking measurements and designing and carrying out experiments. Supplementary teacher materials suggest activities where students work in teams to apply mathematical thinking in making scientific discoveries.

Tom Snyder's programs (1982) also allow for small groups of students working cooperatively or competing against other groups.

Process Functions in Cognitive Technologies

A second set of categories of functions are those which help students understand and use the different mental activities involved in mathematical thinking. Although our understanding of the psychology of mathematics problem solving and learning is continually evolving, there are five different general categories of Process Functions that can be clearly identified for cognitive technologies in math education. Each provides important cognitive support:

- tools for developing conceptual fluency
- tools for mathematical exploration
- tools for integrating different mathematical representations
- tools for learning how to learn
- tools for learning problem-solving methods.

I will briefly define and illustrate each of these categories of functions as they may appear in mathematics software.

1. *Conceptual Fluency Tools.* Fluency tools are programs that free up the component problem-solving *processes* by helping students become more fluent in performing routine mathematical tasks that could be laborious and counterproductive to mathematical thinking. Computer technologies can promote fluency by allowing individually controlled practice on routine tasks and thus freeing up students' mental resources for problem-solving efforts.

There is ample room for debating what these component skills are in secondary school mathematics (e.g. Fey, 1984; Pollak, 1983) and in high schools (Maurer, 1984a, 1984b; Usiskin, 1980). Many software programs routinize irrelevant skills such as practice on long division algorithms. And the issue is made more complex by the fact that there are many mental functions, such as the component operations of numerical and symbolic calculation, that can now be entirely carried out by mathematics software (e.g., Kunkle & Burch, 1984; Pavelle, Rothstein, & Fitch, 1981;

Wilf, 1982; Williams, 1982). We need to focus on determining the skills and knowledge required to design the inputs and understand the outputs of these mathematical tools.

There are routine mathematical tasks that one should be able to do easily to make progress in mathematical thinking. For example, information technologies could improve the fluency of the estimation skills that are at the core of a revised early mathematics curriculum. There are games involving arithmetic estimation activities at which students become quickly proficient (e.g. Pettito, 1984a). The routinization of certain mathematical skills is equally appropriate at the highest level of mathematical achievement. What is at the fringe of mathematical thinking and creativity today is the slog work of tomorrow (Wilder, 1981). What is a creative invention at one point, such as Leibniz's calculus or Gauss's development of complex numbers, is likely to become so routine later that effective instruction makes it widely accessible.

The appropriate roles for such fluency tools is the subject of much current debate (Cole, 1985; Mehan et al., 1985; Patterson & Smith, 1985; Resnick, 1985). Schools frequently establish a two-tiered curriculum, in which basic computational skills are presumed to be necessary prerequisites for engaging in more complex, higher order thinking and problem solving. Limited to activities with little motivational significance in the first tier of this curriculum, many students never engage in the mathematical thinking characteristic of the second tier (Cole, 1985). But recent work in writing (in which such a two-tiered approach is common: Mehan et al., 1985; Simmons, in press) implies that so-called basic skills can be acquired in the *context* of more complex mathematical thinking activities. When a child's conceptual fluency hampers complex thinking, a functional context is established and the child realizes the need for practice. Thus, practice is self-motivated. This contrasts with the two-tiered approach, in which the child is trained to some threshold skill level before being let loose to solve problems. Drill and practice software for fluency in "basic mathematics" seems to work better as a fallback rather than as a startup activity.

2. *Mathematical Exploration Tools.* Mathematics education has long emphasized discovery learning, particularly in the primary school with the use of manipulatives such as Dienes blocks, Cuisenaire rods, and pattern blocks. Much more complex conceptions and mathematical relationships, such as recursion and variables, can also be approached at a more intuitive level of understanding without abstract symbolic equations.

The computational discovery learning environment provides a rich context that helps students broaden their intuition. Logo programming is a paradigm case. The design of Logo environments is based on the assumption that one can recognize patterns and make novel discoveries about properties of mathematical systems through self-initiated search in a well-implemented domain of mathematical primitives (Abelson & DiSessa, 1981; Papert, 1980). Nonetheless, recent findings indicate that students encounter conceptual difficulties—for example, with recursion, procedures, and variables—in Logo and find it hard to understand how Logo dictates flow of control for command execution (Hillel & Samurcay, 1985; Kurland & Pea, 1985; Kurland, Clement, Mawby, & Pea, in press; Kurland, Pea, Clement, & Mawby, in press; Kuspa & Sleeman, 1985; Mawby, in press; Pea, Soloway, & Spohrer, in press; Perkins, Hancock, Hobbs, Martin, & Simmons, in press; Perkins & Martin, 1986). There is consequently much debate about the extent and kinds of structure necessary for successful discovery environments. The question of how well lit the paths of discovery need to be remains open.

Many software programs now offer structured exploration environments to help beginning students over some of these difficulties. Programs such as Delta Draw (Spinnaker) and Turtle Steps (Holt, Rinehart & Winston) are recommended as preliminary activities to off-the-shelf Logo. There are in fact several dozen programs that allow students to use a command language to create designs and explore concepts in plane geometry, such as angle and variable, in systematic ways.

The Geometric Supposers (Sunburst Corporation: for Triangles, Quadrilaterals, and Lines; see Kaput, 1985; Schwartz & Yerushalmy, in press) are striking examples of a new kind of discovery environment. Using these programs, students make conjectures about different mathematical objects in plane geometrical constructions—medians, angles, bisectors. Intended for users from grade 6 and up, it is designed so that students can explore the characteristics of triangles and such concepts as bisector and angle. In this way, students can discover theorems on their own. The program is an electronic straight-edge and compass. It comes with "building" tools for defining and labeling construction parts (like the side of a triangle or an angle) and measurement tools for assessing length of lines, degrees of angles, and areas. Most significant, it will remember a geometric construction the student makes on a specific object (such as an obtuse triangle) as a procedure (as in Logo) and allow the student to "replay" it on new, differently shaped objects (e.g., equilateral triangles). The exciting feature of the environment is that interesting properties that emerge in the course of a construction cry out for testing on other kinds of triangles, and students can follow up. Their task is simplified by the labeling and measurement tools provided for mathematical objects in the construction, and experimentation is encouraged. In fact, several students have discovered previously unknown theorems with the discovery tool. Students find this program an exciting entry into empirical geometry (induction), and it can be used to complement classroom work on proofs (deduction). Kaput (1985) has highlighted the major representational breakthrough in the Supposers: they allow a particular construction to represent a general *type* of construction rather than just itself.

3. *Representational Tools.* These tools help students develop the languages of mathematical thought by linking different representations of mathematical concepts, relationships, and processes. Their goal is to help students understand the precise relationships between different ways of representing mathematical problems and the way in which changes in one representation entail changes in others. The languages of mathematical thought, which become apparent in these different representations, include:

- Natural language description of mathematical relations (e.g., linear equations).
- Equations composed of mathematical symbols (e.g., linear equations).
- Visual Cartesian coordinate graphs of functions in two and three dimensions.
- Graphic representations of objects (e.g., in place-value subtraction, the use of "bins" of objects representing different types of place units).

Mathematics educators have begun to use cognitive technologies in this way as a result of empirical studies demonstrating how competency in mathematical problem solving depends partly on one's ability to think in terms of different representational systems during the problem-solving process. Experts can exploit particular strengths of different representations according to the demands of the problem at hand. For example, many relationships that are unclear in textual descriptions, mathematical equations, or other tables of data values can become obvious in well-designed graphs (Tufte, 1983). One can often gain insight into mathematical relationships, like algebraic functions, by seeing them depicted graphically rather than as symbolic equations (Kaput, 1985, in press).

Interactive technologies provide a means of intertwining *multiple representations* (Dickson, 1985) of mathematical concepts and relationships—like graphs and equations or numbers and pictorial representations of the objects the numbers represent. These representations enhance the symbolic tools available to the student and the flexibility of their use during problem solving. They also have the effect of shortening the time required for mathematical experience by allowing, for example, many more graphs to be plotted per unit of time (Dugdale, 1982), and they make possible new kinds of classroom activities involving data collection, display, and analysis (e.g., Kelman et al., 1983: point and function plotting; histograms).

Manipulable, dynamically linked, and simultaneously displayed representations from different symbol systems (Kaput, 1985) are likely to be of value for learning translation skills between different representational systems, although they are as yet untested in research. For example, students can change the value of a variable in an equation to a new value and observe the consequences of this change on the *shape* of the graph. These experiments can be carried out for algebraic equations and graphs in the motivating context of games like Green Globs (in Graphing Equations by Conduit; cf. Dugdale, 1982) and Algebra Arcade (Brooks-Cole). In these games for grades K–6, the player is given Cartesian x-y coordinate axes with 13 green globs randomly distributed on the graph. The players have to type in equations, which the computer graphs; when a graphed equation hits a glob, it explodes and the player's score increases. Students become skilled at knowing how changes in the values of equations, like adding constants or changing factors (x to 3x, for example), correspond to changes in the shape of the graph (Kaput, personal communication). They discover equation forms for families of graphs, such as ellipses, lines, hyperbolas, and parabolas.

Other examples of dynamically linked representational tools are programs that give visual meaning to operations on algebraic equations (such as adding constants to conditionally equivalent operations). Operations on equations are simultaneously presented with coordinate graphs so that any action on an equation is immediately reflected in the graph shape (Kaput, 1985; Lesh, in press). The student can literally see that doing arithmetic, that is, acting on expressions rather than equations, does not change graphs.

Another example is provided by software programs in which different representations are exploited in relation to one another for learning place-value subtraction. We can point to Arithmekit (Xerox PARC: Sybalsky, Burton, & Brown, 1984), Summit (Bolt, Beranek & Newman: Feurzeig & White, 1984), and Place-Value Place (Interlearn). In each of these cases, number symbols and pictorial representations of objects (and in the case of Summit, synthesized voice) are used in tandem to help students understand how the symbols and the operations on them relate to the corresponding pictorial representations. Only Place-Value Place is commercially available. In this program, a calculator displays the addition and subtraction process using three different representations of number values: a standard number symbol, a position on a number line,, and a set of proportionally sized objects (apples, bushels of apples, crates of bushels, truckloads of crates). As students add or subtract numbers, all three displays change simultaneously to illustrate the operation. Kaput (1985) describes efforts underway at the Harvard Educational Technology Center to develop dynamically linked representational tools for ratio and proportion reasoning activities (e.g., m.p.g. and m.p.h. problems: Schwartz, 1984).

Programs incorporating this function are excellent examples of how the rapid interactivity and representational tools the computer provides create a new kind of learning experience. Students can test out hypotheses, immediately see their effects, and shape their next hypothesis accordingly through many cycles, perhaps through many more cycles than they would with noncomputer technologies.

Computers are also frequently used in displaying graphs and functions in algebra, transformations in geometry, and descriptive statistics. Their use is dynamic and allows student interaction with mathematics in ways that would not be possible in noncomputer environments. In the recent *NCTM Computers and Mathematics Report* (Fey, 1984), the importance of multiple representations in mathematics education is highlighted. In particular, the authors note the rich possibilities for the dynamic study of visual concepts, such as symmetry, projection, transformation, vectors, and for developing an intuitive sense of shape and relationship to number and more formal concepts.

4. *Tools for learning how to learn.* This category refers to software programs that promote reflective learning by doing. They start with the details of specific problem-solving experiences and allow students to consolidate what they have learned in episodes of mathematical thinking. They focus on what both Dewey (1933) and Pólya (1957) describe as the final step in problem solving, reflection that evaluates the work accomplished and assesses the potential for generalizing methods and results (Brown, 1984b). These programs also make possible, in ways to be described, new activities for learning how to learn.

The programs leave traces of the student's problem-solving steps. Tools based on this function provide a more powerful way of learning from experience, because they help students relive their experience. The problem-solving tracks that students leave behind can serve as explicit materials for *studying, monitoring,* and *assessing* partial solutions to a problem as they emerge. They can help students learn to *control* their strategic knowledge and activities during problem-solving episodes (Schoenfeld, 1985a, 1985b).

The crucial feature of such systems is that students have access to trace records of their problem solving *processes* (e.g., the network of steps in geometry proofs: Geometry Tutor [Boyle & Anderson, 1984]; the sequence of operations in algebra equation-solving: AlgebraLand [Xerox PARC; Brown, 1984b]). These records externalize thought processes,

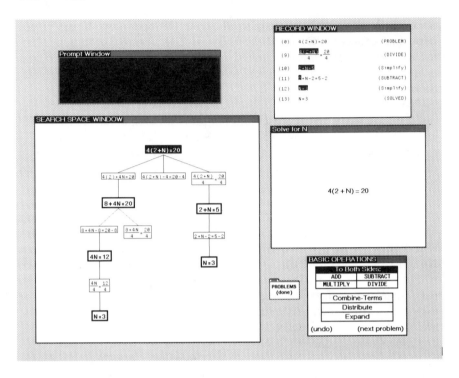

FIG. 4.1.

thus making them accessible for inspection and reflection. Let us consider AlgebraLand, in particular its approach to linear equations.

Search is not a central concept in algebra instruction today, but a central insight of cognitive science is that learning problem-solving skills in mathematics requires well-developed search procedures, that is, knowledge about when to select what subgoals and in what sequence. In most classroom instruction of algebra equation solving, the teacher selects the operator to be applied to an equation (e.g., add-to-both-sides), and the student carries out the arithmetic. The pedagogical flaw in this method is that the student does not learn *when* to select the various subgoals (Simon, 1980), but only *how* to execute them (e.g., to do the arithmetic once the divide operation has been selected).

Originally created several years ago at Xerox PARC by J. S. Brown, K. Roach, and K. VanLehn, and currently being revised by C. Foss for work with middle school students, AlgebraLand is an experimental system for helping students learn algebra by doing problems (Brown, 1984b). Figure 4.1 illustrates some of the features of the system to be discussed.

The task in Fig. 4.1 for the student is to solve the equation for N (shown in the *Solve for N* window on the figure's right side). Algebraic operators listed in the *Basic Operations window* on the bottom right-hand side (such as Combine-Terms, Add-to-Both-Sides, Distribute) can be selected to apply to the whole equation or to one of its subexpressions. After selecting the operation and where to apply it, the student can execute it. This creates a second algebraic expression.

The *Record window,* upper right, records the steps taken by the student towards a solution. Its left-hand column lists all the intermediate expressions; its right-hand column shows each operation used to transform the recorded expression. The *Search Space window* represents all the student's steps as a search tree; it displays solution paths depicting all the student's moves, including backtracking. In the solution attempt depicted here, the student took three different approaches to solving the equation. These approaches are reflected in the three branches that issue from the original equation. Each intermediate expression that resulted from applying the do-arithmetic operator appears in boldface for clarity. AlgebraLand performs all tactical, algebraic operations and arithmetical calculations, effectively eliminating errors in arithmetic or in the application of operators. The student, whose work is limited to selecting the operator and the scope of its operation, is free for the real mental work of search and operator evaluation.

Operators are also provided for exploring solution paths. There is an *Undo* operator that returns the equation to the state immediately preceding the current one and a *Goto* operator, which is not on the menu, that returns the equation to any prior state. The student can also back up a solution path by applying the inverse of a forward operator (e.g., selecting divide just after applying multiply).

Because the windows show every operator used and every state into which the equation was transformed, students have valuable opportunities to learn from their tracks and to *play* with possibilities. They can explore the search paths of their solution space, examining branch points where, on one stem, an operation was used that led down an unsuccessful path, and on another stem, the operation chosen led down a path towards the solution. Then they can decide which features of the equation at the branch point could have led to the best choice. Transforming that decision into an hypothesis, they can test out that hypothesis in future equation solving. These learning activities are *not* possible with traditional methods for learning to solve equations. Indeed, the cognitive technology offered by AlgebraLand affords opportunities for new and different types of learning through problem solving than were available in static, non-computer-based symbolic technologies.

In summary, the computer environment AlgebraLand emphasizes a procedure diametrically opposed to the traditional instructional method. With AlgebraLand, the student decides *when* to apply operators, and the computer carries out the mechanical procedures that transform the equation. Students do face the problem of searching for and discovering a path of operations leading from the problem state to the solution. The graphic representation provided by the search map allows students to reflect on the means they used to solve the problem, and problem solving is no longer an ephemeral process.

AlgebraLand provides a very rich learning and research environment. But we still know far too little about the potential effectiveness of these types of educational instruments, especially since they introduce new learning problems: How do students learn to "read" and use their traces? How do such learning-how-to-learn skills develop, and what ancillary features of software programs such as games will help students understand how to make effective searches in the space of their "thinking tracks"?

The technology is seldom used in this way, and few developers are working along these lines. Nonetheless, when it is used this way, it provides a qualitatively new kind of tool for students to learn more effectively from their problem-solving experience. It can also provide a rich data source for analyzing student understanding of problem-solving processes and methods.

Providing explicit traces of a student's problem solving activities during an episode reveals possible entry points for tutorials in problem-solving skills and for intelligent coaching. In a variant on AlgebraLand, students could be prompted to check the operation (such as factoring) that they have been using during equation-solving activities, for example, if they have used that operator in a way that has led to a more complex equation rather than a simpler one.

Along the same lines, the information available through such traces provides the data for modeling what a student understands; these models could be based on student interactions with a computer in a specific mathematics problem-solving domain. As experience with current prototypes such as the arithmetic game West indicates (Burton & Brown, 1979), such systems allow for coaching dialogues that are sensitive to a student's developmental level and quite subtle in their coaching methods. Unlike intelligent tutoring systems such as the Geometry Tutor (Boyle & Anderson, 1984) and the Lisp Tutor (Anderson & Reiser, 1985), they don't correct the student after every suboptimal move.

There are two problems with using this type of tool for learning how to learn. First, it is not yet apparent how broadly modeling and coaching can be applied to student misconceptions. Second, all the programs that exemplify this category of function run on minicomputers and have not been developed commercially for schools.

5. *Tools for learning problem-solving methods.* This category of tools

encourages *reasoning strategies* for mathematical problem solving. Recent work on the development of mathematical thinking highlights the importance of reasoning strategies. People who have expertise in approaching problem-solving activities in mathematics utilize, in addition to knowledge of mathematical facts and algorithmic procedures, strategies to guide their work on difficult problems that they cannot immediately solve. Such heuristics, well known from the work of Pólya (1957) and modern studies of mathematical problem solving (e.g., Silver, 1985), include drawing diagrams, annotating these diagrams, and exploiting related problems (Schoenfeld, 1985a). Segal, Chipman, and Glaser (1985) review related instructional programs to teach thinking skills.

Few of the existing examples of educational technologies aim to help students develop problem-solving heuristics of this kind. However, one prominent example should be briefly mentioned. Wumpus (Yob, 1975) is a fantasy computer game in which the player must hunt and slay the vicious Wumpus to avoid deadly pitfalls. Goldstein & colleagues (e.g., Goldstein, 1979) created artificial intelligence programs to help students acquire the reasoning strategies in logic, probability, decision analysis, and geometry that are needed for skillful Wumpus hunting. The Wumpus coaches are minicomputer programs that have not been evaluated in educational settings, and the issue of transfer of the skills acquired by students to settings other than this game has not been studied.

Sunburst Corporation (1985) has developed a problem-solving matrix that is indexed to the software it sells for schools, so that purportedly the teacher can know what problem-solving skills and strategies (e.g., binary search) the student will learn by using the program. However, these strategies are not explicitly taught by the software. Furthermore, the scheme erroneously presupposes that one can identify a priori the problem-solving skills that all students will use at all times in working with a specific program. But the problem-solving processes or component thinking skills a student will use in working with a software program change with cognitive development. The skills that are used also vary across individuals because of cognitive style and other variables.

These reservations notwithstanding, if sufficient attention were devoted to the effort, current tools in artificial intelligence could be used to create learning environments in which the application of a mathematics problem-solving heuristic, or set of heuristics, would be exemplified for many problems. Students could explore these applications and then be offered transfer problems that assess whether they have induced how the heuristic works sufficiently to carry on independently (Wittgenstein, 1956). If they have not, the system could offer various levels of coaching, or different layers of hints, that would lead up to a modeling example for that specific problem.

SYNTHESIS OF FUNCTIONS: THE NEED FOR INTERDISCIPLINARY RESEARCH ON COGNITIVE TECHNOLOGIES FOR MATHEMATICS EDUCATION

I would like to close this discussion of functions for cognitive technologies in mathematics education by making a plea for interdisciplinary, classroom-based research, involving mathematics educators, cognitive scientists, software makers, and mathematicians. We know very little about the educational impact, actual or possible, of different technology applications in mathematics education. We cannot predict how the role of the human teacher may change as the use of such technologies increases or in what new ways teachers need to be trained to help students use the technologies to gain control of their own mathematical thinking and learning. As these new technologies reduce the focus on teaching routine computation algorithms, teachers' jobs will become much more intellectually challenging; their activities, more like those of mathematicians. Teacher-training institutions will need to change in as yet unspecified ways. We face a plethora of unknowns. Yet, mathematics educators will have to understand these unknowns in order to use cognitive technologies effectively. Empirical testing of exploratory new instructional curricula that embody the various functions I have described is necessary, for without it, we will have little idea whether the functions actually promote mathematical thinking.

Coda

These are very exciting times for learning mathematics and for using new technologies to shape the futures of mathematical thinking. Mathematics was a dreary subject for many of us in the past, but mathematical thinking is now often learned through problem-solving activities that bury the mechanical aspects of mathematics in interesting ideas. And the puritanical attitude that the mind is a muscle to be exercised through mechanical repetition is giving way to a richer view of the creative, exploring mind, which can be nurtured and guided to discover and learn through meaningful problem-solving activities. By infusing life into the learning tools for mathematics, by integrating supports for the personal side of mathematical thinking with supports for knowledge, we can perhaps help each child realize how the powerful abstractions of mathematics confer personal power. In such a utopia, learning mathematics is but one more way of learning how to think and how to define one's personal voice in the world.

It is hoped that the transcendent functions proposed for cognitive technologies in mathematical education will be useful for crafting new generations of cognitively supportive and personally meaningful learning and teaching tools for mathematical thinking. They may also provide practitioners with new ways of evaluating these educational tools. Other researchers are bound to offer a different set of transcendent functions than those I have proposed; debate should clarify the issues under discussion and contribute to the fundamental goal of using cognitive technologies in mathematics education to prepare students effectively for the complexities of mathematical thinking.

Although we cannot predict what shape the cognitive technologies for mathematics education will take, we can certainly monitor whether they are congruent with emerging concepts of mathematical thinking and the nature of the learner, and assure that the science and tools of education are never far apart.

ACKNOWLEDGMENTS

I have had opportunity to discuss some of these issues with Wally Feurzeig, Ron Mawby, Andrea Pettito, Al Schoenfeld, Judah Schwartz, Kurt VanLehn, Ron Wenger, Jim Wilson, and members of the Sloan workgroup on Cognitive Science and Math Education. Jim Kaput, in particular, has greatly influenced this account of how drastically information technologies are changing and will continue to change our curriculum topics and educational goals in mathematics. None of those mentioned necessarily subscribes to the taxonomy and arguments presented here.

REFERENCES

Abelson, H., & diSessa, A. A. (1981). *Turtle geometry: The computer as medium for exploring mathematics*. Cambridge, MA: MIT Press.

Anderson, J. R., & Reiser, B. (1985). The LISP tutor. *Byte, 10,* 159–175.

Bereiter, C., & Scardamalia, M. (1986). *The psychology of written composition*. Hillsdale, NJ: Lawrence Erlbaum Associates.

Boyle, C. F., & Anderson, J. R. (1984). *Acquisition and automated instruction of geometry proof skills*. Paper presented at the meeting of the American Educational Research Association, New Orleans, LA.

Brown, J. S. (1984a). *Idea amplifiers: New kinds of electronic learning environments*. Palo Alto, CA: Xerox Palo Alto Research Center, Intelligent Systems Laboratory.

Brown, J. S. (1984b). Process versus product: A perspective on tools for communal and informal electronic learning. In S. Newman & E. Poor (Eds.), *Report from the learning lab: Education in the electronic age* (pp. 41–57). New York: Educational Broadcasting Corporation, WNET.

Bruner, J. S. (1966). On cognitive growth II. In J. S. Bruner, R. R. Olver, & P. M. Greenfield (Eds.), *Studies in cognitive growth* (pp. 30–67). New York: Wiley.

Burton, R. R., & Brown, J. S. (1979). An investigation of computer coaching for informal learning activities. *International Journal of Man-Machine Studies, 11,* 5–24.

Cajori, F. (1929a). *A history of mathematical notations, Vol. 1: Notations in elementary mathematics.* La Salle, IL: Open Court.

Cajori, F. (1929b). *A history of mathematical notations, Vol. 2: Notations mainly in higher mathematics.* La Salle, IL: Open Court.

Campione, J. C., & Brown, A. L. (1978). Training general metacognitive skills in retarded children. In M. Grunberg, P. E. Morris, & R. N. Sykes (Eds.), *Practical aspects of memory* (pp. 410–417). New York: Academic Press.

Cassirer, E. (1910). *Substance and function and Einstein's theory of relativity.* Mineola, NY: Dover.

Cassirer, E. (1944). *An essay on man: An introduction to a philosophy of human culture.* New Haven, CT: Yale University Press.

Cassirer, E. (1957). *The philosophy of symbolic forms. Vol. 3: The phenomenology of knowledge.* New Haven, CT: Yale University Press.

Champagne, A. B., & Chaiklin, S. (1985). *Interdisciplinary research in science learning: Great prospects and dim future.* Paper presented at the meeting of the American Educational Research Association, Chicago, IL.

Char, C., Hawkins, J., Wootten, J., Sheingold, K., & Roberts, T. (1983). *"Voyage of the Mimi": Classroom case studies of software, video, and print materials.* New York: Bank Street College of Education, Center for Children and Technology.

Cole, M. (1985). *Non-cognitive factors in education: Subcommittee report.* Washington, DC: National Research Council Commission on Behavioral and Social Sciences and Education.

Cole, M., & Griffin, P. (1980). Cultural amplifiers reconsidered. In D. R. Olson (Ed.), *The social foundations of language and thought: Essays in honor of Jerome S. Bruner* (pp. 343–364). New York: Norton.

Conference Board of the Mathematical Sciences (1983). *The mathematical sciences curriculum K–12: What is still fundamental and what is not.* Washington, DC: CBMS.

Dewey, J. (1933). *How we think.* Chicago: Henry Regnery.

Dewey, J. (1938). *LOGIC: The theory of inquiry.* New York: Henry Holt & Co.

Dickson, P. (1985). Thought-provoking software: Juxtaposing symbol systems. *Educational Researcher, 14,* 30–38.

Dugdale, S. (1982, March). Green Globs: A microcomputer application for the graphing of equations. *Mathematics Teacher,* 208–214.

Dweck, C. S., & Elliot, E. S. (1983). Achievement motivation. In P. H. Mussen (Ed.), *Handbook of child psychology, Vol. IV* (pp. 643–691). New York: Wiley.

Elgarten, G. H., Posamentier, A. S., & Moresh, S. E. (1983). *Using computers in mathematics.* Menlo Park, CA: Addison-Wesley.

Feldman, D. H. (1980). *Beyond universals in cognitive development.* Norwood, NJ: Ablex.

Feurzeig, W., & White, B. (1984). *An articulate instructional system for teaching arithmetic problems.* Cambridge, MA: Bolt, Beranek, & Newman.

Fey, J. T. (1984). (Ed.). *Computing and mathematics: The impact on secondary school curricula.* College Park, MD: National Council of Teachers of Mathematics, The University of Maryland.

Goldstein, I. (1979). The genetic epistemology of rule systems. *International Journal of Man-Machine Studies, 11,* 51–57.

Goodman, N. (1976). *Languages of art.* (Rev. ed.) Amherst, MA: University of Massachusetts Press.

Goody, J. (1977). *The domestication of the savage mind.* New York: Cambridge University Press.

Gould, S. J. (1980). Shade of Lamarck. In *The panda's thumb: More reflections in natural history* (pp. 76–84). New York: Norton.

Grabiner, J. V. (1974). Is mathematical truth time-dependent? *American Mathematical Monthly, 81,* 354–365.

Graves, D. H., & Stuart, V. (1985). *Write from the start: Tapping your child's natural writing ability.* New York: Dutton.

Greenfield, P. (1972). Oral and written language: The consequences for cognitive development in Africa, the United States, and England. *Language and Speech, 15,* 169–178.

Greenfield, P. M., & Bruner, J. S. (1969). Culture and cognitive growth (revised version). In D. A. Goslin (Ed.), *Handbook of socialization theory and research* (pp. 633–657). Chicago: Rand McNally.

Harter, S. (1985). Competence as a dimension of self-evaluation: Toward a comprehensive model of self-worth. In R. Leahy (Ed.), *The development of the self.* New York: Academic Press.

Hawkins, J., Hamolsky, M., & Heide, P. (1983). Paired problem solving in a computer context (*Tech. Rep. No. 33*), New York: Bank Street College, Center for Children and Technology.

Hawkins, J., Sheingold, K., Gearhart, M., & Berger, C. (1982). Microcomputers in schools: Impact on the social life of elementary classrooms. *Journal of Applied Developmental Psychology, 3,* 361–373.

Hillel, J., & Samurcay, R. (1985). *Analysis of a Logo environment for learning the concept of procedures with variable.* (Technical Report). Montreal: Concordia University, Mathematics Department.

Kaput, J. J. (1985). *Information technology and mathematics: Opening new representational windows.* Paper presented at National Conference on Using Computers to Teach Complex Thinking, National Academy of Sciences, Washington, DC.

Kaput, J. J. (in press). Towards a theory of symbol use in mathematics. In C. Janvier (Ed.), *Problems of representation and translation in mathematics.* Hillsdale, NJ: Lawrence Erlbaum Associates.

Kelman, P., Bardige, A., Choate, J., Hanify, G., Richards, J., Roberts, N., Walters, J., & Tornrose, M. K. (1983). *Computers in teaching mathematics.* Reading, MA: Addison-Wesley.

Kessel, F. S., & Siegel, A. W. (Eds.). (1983). *The child and other cultural inventions.* New York: Praeger.

Kessen, W. (1979). The American child and other cultural inventions. *American Psychologist, 34,* 815–820.

Kline, M. (1972). *Mathematical thought from ancient to modern times.* Oxford: Oxford University Press.

Kunkle, D., & Burch, C. I. Jr. (1984). Symbolic computer algebra: The classroom computer takes a quantum leap. *Mathematics Teacher,* 209–214.

Kurland, D. M., Clement, C., Mawby, R., & Pea, R. D. (in press). Mapping the cognitive demands of learning to program. In J. Bishop, J. Lochhead, & D. Perkins (Eds.), *Thinking.* Hillsdale, NJ: Lawrence Erlbaum Associates.

Kurland, D. M., & Pea, R. D. (1985). Children's mental models of recursive Logo programs. *Journal of Educational Computing Research, 1,* 235–243.

Kurland, D. M., Pea, R. D., Clement, C., & Mawby, R. (in press). A study of programming ability and thinking skills in high school students. *Journal of Educational Computing Research.*

Kuspa, L., & Sleeman, D. (1985). *Novice Logo errors.* Unpublished manuscript, Stanford University, School of Education and Department of Computer Science.

Lepper, M. R. (1985). Microcomputers in education: Motivational and social issues. *American Psychologist, 40,* 1–18.

Lesh, R. (1981). Applied mathematical problem solving. *Educational Studies in Mathematics, 12(2),* 235–265.

Lesh, R. (in press). The evolution of problem representations in the presence of powerful cultural amplifiers. In C. Janvier (Ed.), *Problems of representation in mathematics learning and problem solving.* Hillsdale, NJ: Lawrence Erlbaum Associates.

Levin, J. (1984). *Place-value place.* Cardiff-by-the-sea, CA: InterLearn.

Luria, A. R. (1976). *Cognitive development: Its cultural and social foundations.* Cambridge, MA: Harvard University Press.

Luria, A. R. (1979). *The making of mind: A personal account of Soviet psychology.* Cambridge, MA: Harvard University Press.

Mach, E. (1960). *The science of mechanics: A critical and historical account of its development.* 6th ed. LaSalle, IL: Open Court. (Originally published, 1893).

Malone, T. W. (1981). Toward a theory of intrinsically motivating instruction. *Cognitive Science, 4,* 333–369.

Maurer, S. B. (1984a). College entrance mathematics in the year 2000. *Mathematics Teacher, 77,* 422–428.

Maurer, S. B. (1984b). The effects of a new college mathematics curriculum on high school mathematics. In A. Ralston & G. S. Young (Eds.), *The future of college mathematics* (pp. 153–175). New York: Springer-Verlag.

Mawby, R. (in press). Proficiency conditions for the development of thinking skills through programming. In J. Bishop, J. Lochhead, & D. N. Perkins (Eds.), *Thinking.* Hillsdale, NJ: Lawrence Erlbaum Associates.

McLellan, J. A., & Dewey, J. (1895). *The psychology of number: And its applications to methods of teaching arithmetic.* New York: Appleton.

McLeod, D. B. (1985). Affective issues in research on teaching mathematical problem solving. In E. A. Silver (Ed.), *Teaching and learning mathematical problem solving: Multiple research perspectives* (pp. 267–279). Hillsdale, NJ: Lawrence Erlbaum Associates, 267–279.

Mehan, H., Moll, L., & Riel, M. (1985). *A quasi-experiment in guided change.* (#NIE 6-83-0027). Washington, DC: National Institute of Education.

Menninger, K. (1969). *Number words and number symbols: A cultural history of numbers.* Cambridge, MA: MIT Press.

National Science Board Commission on Precollege Education in Mathematics, Science, and Technology. (1983). *Educating Americans for the 21st century.* Washington, DC: National Science Foundation.

Noddings, N. (1985). Small groups as a setting for research on mathematical problem solving. In E. A. Silver (Ed.), *Teaching and learning mathematical problem solving: Multiple research perspectives* (pp. 345–359). Hillsdale, NJ: Lawrence Erlbaum Associates.

Olson, D. R. (1976). Culture, technology and intellect. In L. B. Resnick (Ed.), *The nature of intelligence.* Hillsdale, NJ: Lawrence Erlbaum Associates.

Olson, D. R. (1977). From utterance to text: the bias of language in speech and writing. *Harvard Educational Review, 47,* 257–281.

Olson, D. R. (1985). Computers as tools of the intellect. *Educational Researcher, 14,* 5–8.

Olson, D. R., & Bruner, J. S. (1974). Learning through experience and learning through media. In D. R. Olson (Ed.), *Media and symbols: The forms of expression, communication, and education* (pp. 125–150). Chicago: University of Chicago Press.

Ong, W. J. (1982). *Orality and literacy: The technologizing of the word.* New York: Methuen.

Papert, S. (1972). Teaching children to be mathematicians versus teaching about mathematics. *International Journal for Mathematical Education, Science, and Technology, 3,* 249–262.

Papert, S. (1980). *Mindstorms: Children, computers, and powerful ideas.* New York: Basic Books.

Papert, S., Watt, D., diSessa, A., & Weir, S. (1979). *An assessment and documentation of a children's computer laboratory*. Final Report of the Brookline Logo Project. Cambridge, MA: MIT, Department of Artificial Intelligence.

Patterson, J. H., & Smith, M. S. (1985, September). *The role of computers in higher order thinking*. (#NIE-84-0008). Washington, DC: National Institute of Education.

Pavelle, R., Rothstein, M., & Fitch, J. (1981). Computer algebra. *Scientific American, 245*, 136–152.

Pea, R. D. (1985a). Integrating human and computer intelligence. In E. Klein (Ed.), *New directions for child development: Children and Computers, No. 28* (pp. 75–96). San Francisco: Jossey-Bass.

Pea, R. D. (1985b). Beyond amplification: Using the computer to reorganize mental functioning. *Educational Psychologist, 20*, 167–182.

Pea, R. D., & Kurland, D. M. (1984). On the cognitive effects of learning computer programming. *New Ideas in Psychology, 2*, 137–168.

Pea, R. D., Soloway, E., & Spohrer, J. (in press). The buggy path to the development of programming expertise. *Focus on Learning Problems in Mathematics*.

Perkins, D. N., Hancock, C., Hobbs, R., Martin, F., & Simmons, R. (in press). Conditions of learning in novice programmers. *Journal of Educational Computing Research*.

Perkins, D. N., & Martin, F. (1986). Fragile knowledge and neglected strategies in novice programmers. In E. Soloway & S. Iyengar (Eds.), *Empirical studies of programming* (pp. 213–229). Norwood, NJ: Ablex.

Pettito, A. L. (1984a). *Beyond drill and practice: Estimation and collaboration in a microcomputer environment*. Paper presented at the meeting of the American Educational Research Association, New Orleans, LA.

Pettito, A. L. (1984b). *Discussant's comments on misconceptions and comments on important issues for research in science and mathematics education*. Paper presented at the meeting of the American Educational Research Association, New Orleans, LA.

Pollak, H. 1983). *The mathematics curriculum: What is still fundamental and what is not*. Conference Board of the Mathematical Sciences (H. Pollak, Chair). Washington, DC.

Pólya, G. (1957). *How to solve it*. New York: Doubleday-Anchor.

Resnick, D. P., & Resnick, L. B. (1977). The nature of literacy: An historical explanation. *Harvard Educational Review, 47*, 370–385.

Resnick, L. B. (1985). *Education and learning to think: Subcommittee report*. Washington, DC: National Research Council Commission on Behavioral and Social Sciences and Education.

Rosch, E., & Mervis, C. (1975). Family resemblances: Studies in the internal structure of categories. *Cognitive Psychology, 7*, 573–605.

Schoenfeld, A. (1985a). *Mathematical problem solving*. New York: Academic Press.

Schoenfeld, A. (1985b). Metacognitive and epistemological issues in mathematical understanding. In E. A. Silver (Ed.), *Teaching and learning mathematical problem solving* (pp. 361–379). Hillsdale, NJ: Lawrence Erlbaum Associates.

Schwartz, J. (1984). *An empirical determination of children's word problem difficulties: Two studies and a prospectus for further research* (Tech. Rep.). Cambridge, MA: Harvard Graduate School of Education, Educational Technology Center.

Schwartz, J., & Yerushalmy, M. (1985). *The geometric supposers*. Pleasantville, NY: Sunburst Communications.

Schwartz, J., & Yerushalmy, M. (in press). The geometric supposer: The computer as an intellectual prosthetic for the making of conjectures. *The College Mathematics Journal*.

Scribner, S., & Cole, M. (1981). *The psychology of literacy*. Cambridge, MA: Harvard University Press.

Segal, J. W., Chipman, S. F., & Glaser, R. (Eds.). (1985). *Thinking and learning skills: Relating instruction to basic research (Vol. 1)*. Hillsdale, NJ: Lawrence Erlbaum Associates.

Sheingold, K., Hawkins, J., & Char, C. (1984). "I'm the thinkist, you're the typist": The interaction of technology and the social life of classrooms. *Journal of Social Issues, 40,* 49–61.

Silver, E. A. (Ed.). (1985). *Teaching and learning mathematical problem solving: Multiple research perspectives.* Hillsdale, NJ: Lawrence Erlbaum Associates.

Simmons, W. (in press). Beyond basic literacy skills. In R. D. Pea & K. Sheingold (Eds.), *Mirrors of Minds.* Norwood, NJ: Ablex.

Simon, H. A. (1980). Problem solving and education. In D. Tuma & F. Reif (Eds.), *Problem solving and education* (pp. 81–96). Hillsdale, NJ: Lawrence Erlbaum Associates.

Slavin, R. (1983). *Cooperative learning.* New York: Longman.

Slavin, R., Sharan, S., Kagan, S., Hertz-Lazarowitz, R., Webb, C., & Schmuck, R. (1985). *Learning to cooperate, cooperating to learn.* New York: Plenum.

Sleeman, D., & Brown, J. S. (Eds.). (1982). *Intelligent tutoring systems.* New York: Academic Press.

Snyder, T. (1982). *Search Series.* New York: McGraw-Hill.

Stodolsky, S. (1984). Frameworks for studying instructional processes in peer work-groups. In P. L. Peterson, L. C. Wilkinson, & M. Hallinan (Eds.), *The social context of instruction.* Orlando, FL: Academic Press.

Suppes, P. (Ed.). (1978). *Impact of research on education: Some case studies.* Washington, DC: National Academy of Education.

Sybalsky, R., Burton, R., & Brown, J. S. (1984). *Arithmekit.* Palo Alto, CA: Xerox Palo Alto Research Center.

Tufte, E. R. (1983). *The visual display of quantitative information.* Cheshire, CT: Graphics Press.

Usiskin, Z. (1980). What should not be in the algebra and geometry curricula of average college-bound students? *Mathematics Teacher, 73,* 413–424.

Vygotsky, L. S. (1962). *Thought and language.* Cambridge, MA: MIT Press.

Vygotsky, L. S. (1978). *Mind in society: The development of the higher psychological processes.* Cambridge, MA: Harvard University Press.

Wartofsky, M. (1983). The child's construction of the world and the world's construction of the child: From historical epistemology to historical psychology. In F. S. Kessel & A. W. Sigel (Eds.), *The child and other cultural inventions* (pp. 188–215). New York: Praeger.

Webb, N. M. (1984). Microcomputer learning in small groups: Cognitive requirements and group processes. *J. Educational Psychology, 76,* 1076–1088.

White, S. (1983). Psychology as a moral science. In F. S. Kessel & A. W. Sigel (Eds.), *The child and other cultural inventions* (pp. 1–25). New York: Praeger.

Whitehead, A. N. (1929). *The aims of education.* New York: Macmillan.

Whitehead, A. N. (1948). *An introduction to mathematics.* Oxford: Oxford University Press.

Wilder, R. L. (1981). *Mathematics as a cultural system.* New York: Pergamon Press.

Wilf, H. S. (1982). The disk with the college education. *American Mathematical Monthly, 89,* 4–8.

Williams, G. (1982, October). Software Arts' TK!Solver. *Byte,* 360–376.

Wittgenstein, L. (1956). *Remarks on the foundations of mathematics.* Oxford: Blackwell.

Yob, G. (1975). Hunt the Wumpus. *Creative Computing, 1,* 51–54.

5 Problem Formulating: Where Do Good Problems Come From?

Jeremy Kilpatrick
University of Georgia

If we change the question in the title to Where do good mathematics problems come from?, the answer ought to be readily apparent to any competent high school graduate. Mathematics problems obviously come from mathematics teachers and textbooks, so good mathematics problems must come from good mathematics teachers and good mathematics textbooks. The idea that students themselves can be the source of good mathematics problems has probably not occurred to many students or to many of their teachers.

Problem formulating is an important companion to problem solving. It has received little explicit attention, however, in the mathematics curriculum. Teachers and students alike assume that problems are simply there, like mountains to be climbed, and they seldom ask how problems might be given alternative formulations, let alone where they come from in the first place. Problem formulating should be viewed not only as a *goal* of instruction but also as a *means* of instruction. The experience of discovering and creating one's own mathematics problems ought to be part of every student's education. Instead, it is an experience few students have today—perhaps only if they are candidates for advanced degrees in mathematics.

Students in mathematics classes will soon have access to computer programs that can perform all the routine numerical and algebraic manipulations that have traditionally required months to learn. It is only a matter of time until mathematics instruction starts shifting away from the memorization of standard algorithms toward a more conceptual emphasis on various operations and their use. As that happens, problem formulating ought to become an increasingly important part of the school mathematics curriculum (K. VanLehn, personal communication, December 17, 1984). Furthermore, teachers of programming courses have discovered that they need to refine students' problem-formulating skills.

Researchers, like teachers, have tended to ignore the processes of posing and formulating problems. As Getzels (1979) has observed, "although there are dozens of theoretical statements, hundreds of psychometric instruments, and literally thousands of empirical studies of problem solving, there is hardly any systematic work on problem finding" (p. 167). Most of the little work there is has dealt with problem finding in art (Getzels & Csikszentmihalyi, 1976) and not in mathematics. The cognitive science literature provides us with some ideas on problem formulating that will be discussed later, but most of that literature deals with the reformulation of ill-formulated problems or the formulation of subproblems and related problems. All of that formulation and reformulation is in the service of solving a problem that has already been set for the student. The research may help us understand some problem-formulating processes, but it cannot give a complete picture until more researchers look at the formulation of problems in situations where a problem has not yet been posed.

The purpose of this chapter is to explore the notion of problem formulating in mathematics as it relates to the overlapping interests of mathematics teachers and cognitive scientists. Since the topic has received little systematic study, this chapter is largely part of a ground-clearing operation. It may help define some of the issues that need further thought and investigation before a theoretical edifice can be built. Attempts to teach problem-formulating skills, of course, need not await a theory, and some of the ideas proposed in this chapter may help them along.

SOURCES OF PROBLEMS

Other People as a Problem Source

As noted, almost all of the mathematical problems a student encounters have been proposed, and formulated, by another person—the teacher or the textbook author. In real life outside of school, however, many problems, if not most, must be created or discovered by the solver, who gives the problem an initial formulation.

Psychologists are fond of reminding us that a problem is not a problem for you until you accept it and interpret it as your own. One person cannot give a problem to another person; the second person has to construct the problem for himself or herself. "Problems are not given. They are constructed by human beings in their attempts to make sense of complex and troubling situations" (Schön, 1979, p. 261). In school, however, problems are given.

Wherever the problem comes from, the problem solver is always obliged to reformulate it. In fact, as Duncker (1945) pointed out, one can think of problem solving itself as consisting of successive reformulations of an initial problem. Cognitive science has demonstrated, through computer simulations of problem-solving behavior, that the information given explicitly in a problem statement is almost never adequate for solving the problem. The problem solver has to supply additional information consisting of premises about the problem context. For example, to solve an algebra word problem about the ages of a mother and her son, you need to understand that ages in years cannot be negative numbers. Very often in the case of textbook problems, the additional premises are empirically false. Problems involving a mixture of water and alcohol, for example, may force the solver to make the incorrect assumption that the volume of the mixture is the sum of the volumes of water and alcohol (cf. Davis & Hersh, 1981, pp. 70–74; Simon, 1983, p. 12). The solver needs to supply commonsense knowledge that has not been made explicit and to bridge the gap between a mathematical formulation and the world assumed in the problem statement (Simon, 1983).

The need to know what is not stated in the problem begins in the earliest grades. De Corte, Vershaffel, and De Win (1985) noticed that textbook problems on simple addition and subtraction often presuppose an understanding of unstated relations among the quantities in the problem. Consider, for example, the problem, "Ann and Tom have 8 books altogether. Ann has 5 books. How many books does Tom have?" The statement does not say explicitly that Ann's five books are part of the eight they have together; you are supposed to make that inference yourself. When De Corte et al. reworded problems so that the semantic relations were made more explicit—for example, changing the second sentence of the above problem to read, "Five of these books belong to Ann"—first and second graders made fewer errors in their solutions, and their errors suggested that they had a better understanding of the reworded problems. De Corte et al. contended that teachers and textbook writers should pay more attention to the wording of verbal problems, using rewordings that might help children overcome some of the difficulties they may be having in interpreting a problem.

In solving a practical problem involving mathematics, the need for reformulation may be even clearer. Consider the problem given in Fig. 5.1, which was used in a unit on problem formulating prepared by the School Mathematics Study Group. The original problem appears reasona-

Tom wants to make a clothes-drying rack
for the backyard. He can fix it so that
the clothes lines are strung between two
supports as in Figure 1a or between cross-
bars as in Figure 1b.

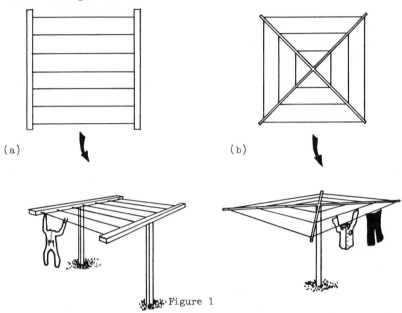

(a)   (b)

Figure 1

How many feet of clothesline would he have to get for each of these
choices if the outer square measures 6 feet on a side and the sepa-
ration between adjacent lines is 1 foot?

Fig. 5.1.   Sample problem. From School Mathematics Study Group (1970), p. 52.

bly well defined, like almost all textbook problems, but it can be used by
teachers as the basis for a discussion on reformulating problems.

Someone named Tom wants to make a clothes-drying rack, and he has
two ways to string the clotheslines. Which way should he do it? The
problem gives students an opportunity to develop and refine mathematical models. The two choices can be modeled initially as seven parallel
lines (*a*) or three concentric squares (*b*) (Fig. 5.2). These models yield $7 \times 6 = 42$ feet for *a* and $4(2 + 4 + 6) = 48$ feet for *b*. But both models fall
short of a realistic solution. Neither model includes the clothesline that
would be needed to tie the ropes to the supports so that they would
actually stay up. The initial mathematical model has simplified matters a
bit too much. So we might try a second model. Suppose each knot

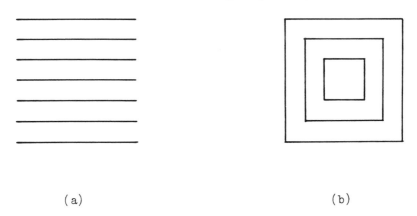

(a)   (b)

Fig. 5.2.   Initial mathematical models. From School Mathematics Study Group (1970), p. 53.

requires an extra 3 inches of clothesline. Then we could represent the knots Tom would need, count them, and add the lengths to the lengths we had previously. (The heavy dots in Fig. 5.3 represent the knots.) We find that $42 + 3\frac{1}{2} = 45\frac{1}{2}$ feet of line are needed for a, and $48 + 6 = 54$ for b. The new model amounts to a reformulation of the problem because we are modeling the situation in a more realistic way.

At this point, the class might want to go further and consider an alternative solution. Maybe we could get along without all the separate lengths—which entail a lot of measuring and cutting. Could we solve the problem using a single length of clothesline for each pattern? Suppose the clothesline were strung in a continuous fashion, as shown in Fig. 5.4. Then, with allowance for the knots, $42 + 6 + \sqrt{\frac{1}{2}} = 48\frac{1}{2}$ feet would be needed for a and $48 + 2\sqrt{2} + \frac{1}{2} \approx 51.3$ feet would be needed for b.

The theme of the class's subsequent discussion might be not only that the solution is being refined but also that the problem itself is being reformulated. One can observe two kinds of reformulation. The first involves refining the mathematical model that was constructed for a situation. The second involves posing the problem in an alternative way that may lead to a different mathematical model. The underlying issue in either case is how reformulation changes the problem.

Other Problems as a Problem Source

As the clothesline problem illustrates, problems can themselves be the source of new problems. There seem to be two phases in the solution process during which new problems can be created. As a mathematical model is being constructed for a problem, the solver can intentionally change some or all of the problem conditions to see what new problem might result. And after a problem has been solved, the solver can look

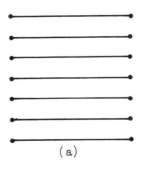
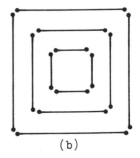

Fig. 5.3. Models with extra line for knotting. From School Mathematics Study Group (1970), p. 53.

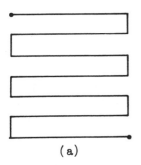
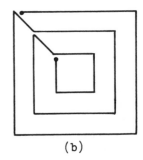

Fig. 5.4. Models with continuous line and knotting. From School Mathematics Study Group (1970), p. 54.

back to see how the solution might be affected by various modifications in the problem.

Making a plan. In making a plan to solve a given problem, a student may take Pólya's (1957) heuristic advice to see whether, by modifying the conditions in the problem, a new, more accessible problem might result that could be used as a stepping stone to solve the original one. An example is the problem of finding the sum of the consecutive odd numbers from 1 to 2001. By changing 2001 successively to 3, 5, 7, 9, and then $2n +$ 1—each time posing a different problem—the solver can discover a pattern and make a generalization that can yield the solution to the original problem. The simpler problems become stepping stones that lead to a final solution. Another example is the problem of finding the incenter of a triangle. The problem can be broken into the two problems of finding the set of points equidistant from two sides of the triangle and finding the set of points equidistant from two other sides of the triangle. In this case, the problem has been reformulated into two simpler subproblems.

As noted earlier, problems are continually being reformulated as they are being solved. "What typically happens in a prolonged investigation is that problem formulation and problem solution go hand in hand, each eliciting the other as the investigation progresses" (Davis, 1985, p. 23). The newly formulated problem may be only a stepping stone to the solution of the original problem, or it may turn out to be a problem of considerable interest in its own right.

Research on the design of software programs suggests that the process of decomposing the design problem into more manageable subproblems is central to the task and that what distinguishes experts from novices is the mastery of this decomposition. Jeffries, Turner, Polson, and Atwood (1981) asked four experienced software designers and five undergraduates who were moderately experienced in programming but not in software design to design a book-indexing system that would provide a list of specified index terms from the text of the book, together with the page numbers on which each term appeared. Both the experts and the novices tended to formulate an overall model of their solution before they began and then to decompose the problem into subproblems. The novices, however, tended to examine fewer subproblems, had more trouble in solving the subproblems they did examine, and seemed to lack ways of effectively representing the knowledge they needed to solve the subproblems.

One novice noticed early in the process of designing the indexing system that a term composed of several words might straddle consecutive pages of the book. He decided that this possibility made the problem unnecessarily complicated, so he ruled it out, making an assumption that he noted in writing. Later he noticed the possibility once again. Apparently not recalling his earlier decision, he treated the second observation as a new discovery. This time, he decided to allow terms to straddle page boundaries, but in so doing he assigned the term the second page number rather than the first. He wrote down this decision, too, never noticing that it contradicted his earlier one.

Jeffries et al. (1981) concluded that novices labor under two handicaps:

> First, the solution to these problems consumes such a large portion of their resources that they are unable to monitor memory for other potentially relevant information. . . . Second, their memory representation of the problem is not organized in such a way as to facilitate the retrieval of previously generated information. (p. 277)

Novices often get mired in the details of solving a particular subproblem without being able to anticipate where it might lead or how it might interact with other subproblems. The expert software designer not only formulates an appropriate set of subproblems but also has a repertoire of methods available to create programs for solving specific subproblems.

The availability of these methods frees the expert to take a wider view of the problem and to anticipate the potential interaction of subproblems. For example, there are different ways of storing the list of terms for an index, and the data structure you use to store the list will constrain what you are able to do in other parts of your program. In more general terms, your solution to one subproblem can influence the nature of the other subproblems that you have identified.

The interaction between subproblems has been studied by researchers in artificial intelligence who are concerned with hierarchical planning. One kind of artificial intelligence program that can plan the solution to a simple problem is Earl Sacerdoti's NOAH (Nets of Action Hierarchies) program (Cohen & Feigenbaum, 1982, pp. 541–550). The program uses a set of procedures termed *critics* that are sensitive to the effects of actions taken in solving a subproblem that might jeopardize the success of the solution plan. The critics allow NOAH to resolve the dilemma of interacting subproblems in a constructive way: the program does not order the operators needed to solve a problem until a potential interaction is found, and then an ordering is chosen that avoids the interaction.

A second method of resolving the dilemma is used by another kind of artificial intelligence program, Mark Stefik's MOLGEN program (Cohen & Feigenbaum, 1982, pp. 551–556), which plans experiments in molecular genetics. MOLGEN has a layered control structure, so that "plans are constructed in one layer, decisions about the design of the plan are made in a higher layer, and strategies that dictate the design decisions are made at a still higher level" (p. 551). The program makes interactions between subproblems explicit, representing them as *constraints* so that it can find a strategy to avoid their potentially harmful effects. For example, when the subproblem of sorting two types of bacteria has been solved by deciding to eliminate one type with an antibiotic, the subproblem of choosing bacteria and antibiotics for an experiment would then be constrained by the requirement that one type of bacteria needs to be susceptible to one of the antibiotics. MOLGEN is able to undo one of its decisions that would lead to an impasse, but it usually successfully avoids the need to backtrack by deferring decisions for which it lacks constraints. Stefik's work suggests that the development of a complex plan requires planning on several levels of abstraction at once.

The cognitive science literature suggests that in devising a plan to solve a problem, successful solvers will distance themselves from the details so that they can see how to reformulate the problem to avoid unmanageable or interacting subproblems. A question for teachers and researchers to ponder is whether, by drawing students' attention to the reformulating process and giving them practice in it, we can improve their problem-solving performance.

Looking back. After a problem has been solved, the solution or the problem itself can suggest additional problems. For example, when you have found which positive numbers can be expressed as the sum of consecutive positive numbers (Mason, Burton, & Stacey, 1982, p. 76), you can ask in how many different ways a given number can be so expressed. The two problems can then be extended to sums of squares or sums of consecutive odd numbers. Pólya (1957) discussed looking back at a solution, and Brown and Walter (1983) suggested ways (to be discussed) in which already solved problems can serve as a source for new problems.

You can also "look back" at a problem that you have solved incompletely or incorrectly to see if formulating it another way might lead you to a better solution. Computer scientists find that searching for the bugs in a program often leads them to reformulate the original problem. A similar approach—viewing one's errors in solving a mathematical problem as bugs that provide an opportunity for debugging—may be helpful to students. The problems and benefits of treating errors as bugs are discussed by Steve Maurer in chapter 7.

Situations as a Problem Source

Mathematical problems often arise in the pursuit of other problems, but they may just as easily come out of the exploration of a situation that has been given a mathematical formulation. People who want to apply their mathematical knowledge to solve a real-life problem find that the formulation of the problem in terms that will permit a mathematical solution is typically a more challenging and creative endeavor than obtaining the solution itself. School exercises in constructing mathematical models of a situation presented by the teacher are intended to provide students with experience in formulating problems. The situation, however, may be perceived by the students as rather contrived. Much work remains to be done in studying the conditions under which students can come to accept a problematic situation as presenting a legitimate challenge to their mathematical abilities rather than as simply another artificial school task. The Committee on the Undergraduate Program in Mathematics (1981, ch. 5) collected useful ideas on mathematical modeling for college students, and Burkhardt (1981) discussed ways that teachers can incorporate modeling into school mathematics courses.

"A problem never comes out of the blue. It is always related to our background knowledge" (Lakatos, 1976, p. 70). Mathematical problems can arise as conjectures when we notice a pattern. For example, we may notice that the amount of crust on a pizza (measured as its circumference) is a linear function of its diameter, the amount of filling (measured as its area) is a quadratic function, and the price is neither (College Board, 1985, pp. 46–48). The problem of comparing the crusts, fillings, and prices of two 9-inch pizzas to those of one 18-inch pizza may then become easier to grasp, and similar problems easier to formulate.

Bundy (1983, ch. 13) has discussed how the Automated Mathematician (AM) program, written by Lenat, uses what is essentially a concept formation process—looking at regularities in a series of examples of a concept—to develop interesting mathematical conjectures. For example, AM discovered the concept of prime by looking at divisors of various numbers, noticing that the number of divisors varies, considering some numbers that have very few divisors, deciding to look at extreme cases, deciding that numbers with 0 or 1 divisor are likely nonexistent and therefore uninteresting, noticing that numbers with 3 divisors seem always to be squares of numbers with 2 divisors, and deciding that numbers with 2 divisors are therefore of special interest. After examining additional examples of primes and factorizations of numbers into primes and following the heuristic rule of asking whether the relation between a number and its divisors is a function, AM conjectured the fundamental theorem of arithmetic, namely, that any integer greater than 1 can be expressed as a product of primes in essentially only one way. Among AM's other interesting conjectures are de Morgan's laws, Goldbach's conjecture, and some theorems on maximally divisible numbers (integers with more divisors than any smaller positive integer) that were discovered by Ramanujan.

The AM program turned out to be less successful in discovering powerful domain-specific heuristic rules in areas other than mathematics. When Lenat attempted to get AM to take heuristics itself as a task domain in which to explore, observe, define new concepts, and develop new conjectures, the results were disappointing. Subsequent work on a descendent of AM called EURISKO (Lenat, 1983) led not only to a program that could discover new heuristic rules but also to some suggestions about why AM had done so well in discovering mathematical concepts and so poorly in heuristics. Lenat observed that AM was successful in uncovering interesting mathematical concepts because it was coded in Lisp, a language whose structure has strong ties to mathematics. EURISKO was successful in its domain partly because it was coded in a new language that permitted a natural and compact statement of heuristic rules. Furthermore, part of the interesting quality of AM's discoveries in number theory turned out, on careful reexamination, to have been attributed to them by human observers (Lenat & Brown, 1983).

Lenat (1983) has argued that to discover interesting information about a concept, one needs heuristic rules that are especially suited to the domain of that concept. It also appears that the form in which one's knowledge—both conceptual and procedural—is represented plays a central role in determining one's ability to acquire those heuristic rules. The processes of concept formation and rule acquisition seem to be interactive. Although one can question the relevance for mathematics instruction of much of the work in artificial intelligence, Lenat's work clearly suggests that teachers should not postpone opportunities for students to develop heuristic rules by learning to formulate problems. The unstated principle behind much mathematics instruction is that students need to develop concepts before they can solve problems, let alone formulate them. The principle that emerges from the experience with AM and EURISKO is that concept formation, problem solving, and problem formulation ought to proceed together.

If students are to become autonomous problem formulators, they will need practice in adopting a more reflective approach to problematic situations. Mathematics educators have begun to suggest ways in which such an approach might be fostered (see Kilpatrick, 1985, for a discussion of some suggestions). In particular, Mason et al. (1982, ch. 8) observe that learning to solve problems entails becoming your own questioner, which is part of the attitude of thinking mathematically. They note that typical mathematical questions are of the following form:

How many . . .?
In how many ways . . . ?
What is the most/least . . . ?
What properties does . . . have?
What is the same about (several different events, facts, situations)?
Where have I seen something like this before?
What is the essential idea here?
What makes this work? (pp. 165–166)

These are questions that are useful not only in solving a problem but also in formulating a problem out of a given situation.

PROBLEM STRUCTURE

Simon (1973) has distinguished between well-structured and ill-structured problems. Well-structured problems are the kinds you ordinarily encounter in school or in a puzzle book. The pertinent information needed to solve the problem is contained in its statement, the rules for finding a correct solution are clear, and you have definite criteria for a solution. Ill-structured problems, on the other hand, are more like the problems you encounter in daily life. Examples include whether you should take out a loan to finance an addition to your house and how to set up a system for indexing a book. Not every problem in pure mathematics is well structured, as Davis (1985) demonstrates, but virtually every problem in applied mathematics is initially ill structured. Students are seldom exposed to ill-structured problems in mathematics classes.

The statement of an ill-structured problem may not contain the information needed to solve it. There may be too much information or not enough; a problem statement may lack sufficient information to yield a unique solution or may contain extraneous or contradictory information. The problem solver must then reformulate the problem by either providing or removing information.

At any step in the solution of an ill-structured problem, the options for the next step may be unclear. It may not even be obvious when a solution has been reached. According to Simon (1973), well-structured and ill-structured problems probably require the same problem-solving processes. With ill-structured problems, however, the solver's concept of the problem changes markedly, and those changes demand a wide repertoire of processes for recognizing a solution.

Fredericksen (1984) has suggested that the dichotomy between well-structured and ill-structured problems is too simple to capture the variations in problem structure. He distinguishes three categories of problems:

1. *Well-structured problems* are clearly formulated, can be solved by the application of a known algorithm, and have criteria available for testing the correctness of a solution. An example is the problem of finding the area of a triangle given the lengths of its sides.

2. *Structured problems requiring productive thinking* are similar to well-structured problems, but the problem solver must devise all or part of the solution procedure. An example is the problem of proving that if a quadrilateral has one pair of opposite sides that are parallel and equal, the other two sides are equal also. The usual proof requires the introduction of a diagonal line that is not given. Furthermore, for this theorem, as for most reasonably complicated theorems, no algorithm is available to provide a proof.

3. *Ill-structured problems* lack a clear formulation, a procedure that will guarantee a solution, and criteria for determining when a solution has been achieved. An example is the problem of finding all the paths from my house to my school. Structuring the problem would require specifying what is to count as a path. Solving it would require developing a procedure for identifying and enumerating the paths.

Both Simon's (1973) and Fredericksen's (1984) classifications of problems blur the distinction between problems in which information for a complete formulation is lacking and problems in which, given a complete formulation, a solution procedure is lacking. Other distinctions can be drawn. For example, one can distinguish between the states of one's knowledge about a problem, as Davis's (1985) characterization of 26 such states compellingly shows.

Greeno (1980) goes further than Simon and argues that there is evidence for the claim that the processes for solving problems requiring productive thinking are not essentially different from the processes for solving well-structured problems. Greeno, Magone, and Chaiklin (1979) used a problem-solving simulation model called Perdix to study some elementary geometric constructions. They found that when Perdix was faced with the problem of proving a theorem for which a construction was needed, the solution appeared to depend on planning knowledge organized in a hierarchical fashion, as it does in NOAH or MOLGEN. For example, to prove the pons asinorum—that the base angles of an isosceles triangle are equal—Perdix, like the students to whom the problem was given, constructed the bisector of the vertex angle and proved the resulting triangles congruent by the side-angle-side postulate. Greeno argues that for problems in which there is insufficient information in the initial problem space, new problem spaces must be generated during the course of problem solving. Perdix uses planning knowledge to identify the new elements that might be needed for a solution, and Greeno sees that as evidence for viewing the solving of structured problems requiring productive thinking as simply an extension of the solving of well-structured problems.

The pons asinorum may not be as poorly structured as Greeno's view suggests. A theorem-proving machine developed by Gelernter found a less well-known but much more elegant proof of the pons asinorum that involved self-congruence of the given triangle and did not require the introduction of a construction line (McCorduck, 1979, p. 187). The moral is that the structure of a problem may be in the eye of the problem solver.

PROCESSES OF PROBLEM FORMULATING

Problem finding occurs partly as a result of a generally creative disposition and partly from a habit specifically of problem finding. . . . A pattern of exploration in the early stages of narrowing down and readiness to revise early decisions in the later stages is characteristic of creating. The pattern seems to be part symptom and part cause of inventiveness. (Perkins, 1981, p. 187)

Can the creative disposition and the habit of problem finding be engendered in our students? The answer may depend on achieving a better understanding of the processes of problem formulating. Although the processes needed to solve a poorly structured problem may be little more than extensions of the processes needed to solve a well-structured problem (Greeno, 1980; Simon, 1973), the processes needed to formulate an original problem of either sort may be rather different from the processes needed to reformulate that problem, regardless of how it has been posed. The following paragraphs are offered as preliminary thoughts, not as a comprehensive treatise, on the shape some of the problem-formulating processes might take.

Association

Cognitive scientists have proposed a variety of models of human knowledge, all of which characterize knowledge as structured and organized (Calfee, 1981; Resnick & Ford, 1981, ch. 8). In general, the models tend to portray items of knowledge as arranged in a hierarchical network structure. If knowledge is represented as a network of associated ideas, that network can be used to generate problems by taking a concept node in the network and raising questions about its associates.

Novak and Gowin (1984) advocate the use of *concept maps* to represent the organization of concepts. A concept map is a network of concept labels joined by linking words to form propositions that express relations between the concepts. For example, the concepts *triangle* and *isosceles* might be represented in a concept map by two labeled nodes connected by a line labeled *can be*. The simple proposition thus represented is that a triangle can be isosceles. Novak and Gowin argue that a concept map provides the teacher with a road map for guiding learning so as to honor the meanings of concepts as they are used in propositions. To aid learning, a concept map ought to be organized in a hierarchy, with the most general and inclusive concepts at the top of the map and the most specific and restricted concepts at the bottom. The activity of constructing a concept map, according to Novak and Gowin, is a great aid in reflecting on one's understanding of those concepts and may help foster creative thinking about them.

Brown and Walter (1983) suggest some activities, which they label "accepting the given," that might be helpful in using a concept map to generate problems. For example, one's concept of *isosceles triangle* might be associated in a network with the concepts of *Greek term, right triangle,* and *symmetric,* among others. From those associations, one might then derive questions such as the following: (a) Why is it called isosceles? (b) What properties of isosceles right triangles are not shared by other right triangles? By other isosceles triangles? and (c) What types of symmetry does an isosceles triangle have? Concept mapping is a process for making explicit one's associations with a concept, and accepting the given is a process for using those associations to pose new problems about the concept. Together they might provide mathematics teachers with a productive way of treating problem posing as a school activity.

Analogy

"The key instrument of the creative imagination is analogy" (Wilson, 1984, p. 454). Pólya (1954) showed that analogy can be a fertile source of new problems. For example, after establishing the Pythagorean proposition in the plane—namely, that in a right triangle the sum of the squares of the legs equals the square of the hypotenuse—you can ask what the analogous proposition might be in solid geometry. Is there more than one such analogy (Pólya, 1981, vol. 1, pp. 34–37)?

Not every analogy is productive, however. The key, as in any creative activity, is selection, discernment—the ability to see below the surface of a fuzzy notion of similarity to the relationship that can yield new problems for investigation. "It may well be that it is precisely in the process of turning an initially vague, rich, multi-purpose feeling of analogy into a well-clarified model that much of the creative process in science takes place" (Gentner, 1982, pp. 128–129). The teacher's task, for which cognitive science as yet has little to offer, is to help students develop an attitude toward problem posing that encourages them to seek out analogies and sensitizes them to those analogies that might be productive of interesting new problems.

Generalization

Cognitive science has attempted with some success to model the human process of inducing a generalization from several instances. For example, Anderson, Kline, and Beasley (1979) developed a computer simulation called ACT (Adaptive Control of Thought) that needed only two instances to produce a generalization. In an interesting aside, they observe that generalizing over two instances has a certain logic: "Suppose one person comes into your office and says, 'I cannot make our appointment. I am going to Brazil.' A second person comes into your office and says, 'Could you teach my class for me, I am going to Brazil.' You immediately ask the question, 'Why is everyone going to Brazil?' " (Anderson, Kline, & Beasley, 1979, p. 287).

What is not often noticed is that in mathematics, generalizations can be made from one instance. For example, a gifted 11-year-old, when asked to solve the "unrealistic" problem of representing the general form of numbers that leave a remainder of 7 when divided by 5, said, "In general, in these cases . . . we must take the multiplier x for the given number and add the remainder so that it is divisible by y and there will be z in the remainder. . . . All the numbers will be: $xy + z$. In the given case it will be $x \cdot 5 = 7$" (Krutetskii, 1976, p. 207). The ability to generalize from a single case appears to be a characteristic of students who are talented in mathematics.

Generalization and its companion, specialization, can be a source not only of problem solutions but of new problems as well (Pólya, 1981, Vol. 2, p. 51). The question of how generalization from one or more instances yields a new problem, however, remains virtually unexplored in either research studies or practical experience.

Contradiction

Brown and Walter (1983) have identified a powerful problem-posing strategy termed *what-if-not* that generates problems by contradicting one or more parts of an assertion. For example, you might begin by describing the Pythagorean theorem with a list such as:

1. The statement is a theorem.
2. The theorem deals with a right triangle.
3. The theorem deals with areas.
4. The variables are related by an equals sign.

The four statements can be termed *attributes* of the theorem. Other attributes might be added. You then confront each attribute with one or more what-if-not questions: What if the statement were not a theorem? What else could it be? What if it were a false statement? What if the theorem did not deal with right triangles? What if it dealt instead with an acute triangle? With an obtuse triangle? What if it did not deal with triangles at all? What if it dealt with volume instead of area? What if the relation were not equality but inequality? And so on.

Variants of the what-if-not strategy are possible. Jim Kaput (personal communication, December 17, 1984) has suggested a *what-if-more* approach: What if the exponents in the algebraic formulation of the Pythagorean theorem were not 2, but more than 2? What if the number of terms were more than 3? Another possibility is to take something that is a constant in a problem and allow it to vary. What happens to the Pythagorean theorem if the sum of the angles in the triangle varies above or below 180 degrees?

In each case, the what-if-not strategy gives rise to a variety of new problems that can be explored for their consequences. Brown (1984) has suggested that the stages of the strategy do not lend themselves to mechanization: "The process tends to elude a computerized mentality" (p. 20). It should, however, be open to some systematic investigation.

Other Processes

One process that might be helpful to students in learning to formulate problems is to attempt to take another person's point of view: "How would X think about this problem?" "What might have been responsible for getting people to think about this issue?" (See Brown, 1984, p. 19, on the uses of "pseudo-history.")

Another process for generating ideas for problems is taking characteristics of two concepts and forming their Cartesian product or their intersection. For example, what features of isosceles triangles might be associated with those of 2-by-2 matrices? What do complex numbers and circles have in common? Such fanciful questions may help students discern relationships they did not see before and may yield material for further investigation.

The literature on problem representation may yield additional processes for generating problems (see Jim Greeno's discussion of representation in chapter 3). How a problematic situation is represented will determine what problems can be derived from the situation as well as how they are solved once derived.

INSTRUCTION IN PROBLEM FORMULATING

How to design instruction that will help students learn to formulate mathematical problems is itself a problem in need of a more complete formulation. In the absence of the theory, research, and experience needed to reveal various facets of the problem, only the most tentative conjectures can be advanced.

Merely immersing students in an environment rich in mathematical problems, with the expectation that they will catch on to how good problems are constructed, seems unlikely to improve their problem formulations. The environment needs other forms of enrichment.

The computer is a valuable tool for exploring problematic situations. It should be available in the mathematics classroom along with a variety of other materials for exploration. Students can use the computer to vary the data in a problem and see how those variations affect the solution. With the computer they can generate number patterns that may yield conjectures they can test and try to prove. They can use graphical displays to suggest plausible theorems in geometry. They can easily create new problems by varying the syntax of a problem statement and study the difficulties their peers have with the revised problems. The computer supports a wide range of exploratory activities.

The teacher, however, is the key part of the enriched classroom environment. To engage in the creative act of problem formulating, students need a climate in which the exploration of ideas is encouraged (Frederikson, 1984; Getzels, 1964). That climate will not develop unless the teacher can actively promote it. Creating such a climate may entail restructuring the reward system in the classroom so that teachers and students alike can profit from the students' problem-formulating activities. At present, there is little opportunity and little reward for such activities.

Recently, some children from kindergarten to grade 6 were confronted with "problems" in which no question was posed:

> Mr. Lorenz and 3 colleagues started at Bielefeld at 9 am and drove
> the 360 km to Frankfurt, with a rest stop of 30 minutes.

These stories were inserted into a set of ordinary word problems. The higher the grade level, the more likely the children were to attempt a calculation to solve the problems. Very few children in kindergarten or grade 1 used the story data in a calculation, whereas most of the children in grades 3 to 6 obtained a solution (Radatz, cited in Kilpatrick & Radatz, 1983). This finding suggests that schooling may encourage students to see school problems as detached from the real world of problems. It also suggests that mathematics teachers can pose problems without questions to their students as a means of investigating how deeply they have internalized the view that school problems represent a special form of reality. (See also Krutetskii, 1976.)

In a study that sought to assess the degree of abstraction in the questions posed by children given a set of objects such as scissors, candles, and thumb tacks, Arlin (1977) found, with children in grades 2, 4, and 6, that the older children posed more abstract questions than the younger children did. Below the level of formal operations, general questions did not appear. The quality of the questions children pose may be an index of both their approach to problem finding and how well they can solve problems. That same inference can be drawn from research discussed by Greeno in chapter 3 on the differences between experts and novices in how they represent problems. As Miyake and Norman (1979) have shown, novice learners are often unable to pose questions about a topic because they know too little about it, whereas learners who know a great deal about a topic may have few questions to raise. In learning about a topic, questions arise when the learner knows enough but not too much; the relation between questioning and knowledge appears curvilinear. In formulating problems, however, additional knowledge may well lead increasingly to a greater questioning ability.

There seems to have been little research on the effects of giving students problems to reformulate or asking them to create their own problems. The research that is available suggests that the effects on problem-solving performance are positive (Cohen & Stover, 1981; Keil, 1965; Stover, 1982). Stover taught sixth graders to modify problems in one of three ways: adding a diagram, removing extraneous information, or reordering the information in the problem. Instruction in each type of modification led to a striking improvement in the students' ability to solve problems that were suited to that modification. Keil found that sixth graders who had experience in writing and solving their own problems in response to a situation did better on subsequent tests of achievement in mathematics than a comparable group who solved textbook-type problems about the situation. In these studies, unfortunately, the effects of the experimental treatments on the students' ability to formulate problems were not assessed.

Questions asked by Frederiksen (1984) about instruction in problem solving can be asked as well about instruction in problem formulating: Should problem-formulating skills be taught explicitly or by discovery? How general should instruction in problem formulating be? Furthermore, one can ask how instruction in problem formulating might interact with instruction in problem solving.

Teachers who wish to explore such questions might find Caldwell's (1984) sample unit plans helpful. The two plans illustrate how problem characteristics can be varied and new problems constructed from old ones. For example, a teacher might begin by demonstrating how the order of the information given in a problem can be changed without changing the nature of the problem itself. The students would then be given a series of increasingly complex problems to rewrite. In subsequent lessons, students might be given a situation (e.g., prices of different-sized boxes of soap in a supermarket) and asked to devise problems arising in the situation that made use of specified number facts.

Another avenue for exploration is the mathematical investigation—an extensive assignment that takes some time to complete and that may require the efforts of several students working together. Such investigations have been promoted in the United Kingdom (Branca, 1985). They have the obvious advantage of yielding a variety of subproblems that require formulation before they can be addressed.

Group work seems to provide a natural context for problem formulating. When students work together, they often identify problems that would be missed if they were working alone. A poorly formulated idea brought up by one student can be tossed around the group and reformulated to yield a fruitful problem. Students participate in a dialogue with others that mirrors the kind of internal dialogue that good problem formulators appear to have with themselves. Collaborative work apparently helps students learn to control their problem-solving performance (Schoenfeld, 1985, pp. 140–143); it would probably help them improve their problem-formulating performance as well. Students need time to reflect on the problem-formulating process, and they need to develop a language for talking with one another and with the teacher. Compared to a lecture-discussion class, a class organized so that the students can work together on mathematical investigations appears to provide a richer context for reflection and language development.

UNDERSTANDING AND DEVELOPING PROBLEM-FORMULATING ABILITIES

"A fool will ask more questions than the wisest can answer."
—Jonathan Swift

A host of issues can be raised concerning prospective research on understanding and developing problem-formulating abilities. However, just as it is easy to pose many problems and hard to pose good problems, so it is easy to raise many issues and hard to raise those issues that will turn out to be fruitful for teachers and researchers. Some issues that appear to have promise concern (a) the effects on problem-solving performance of instruction and practice in problem formulating, (b) the differences between expert and novice problem solvers in their problem-formulating processes, and (c) the quality of problems formulated as a measure of mathematical creativity.

Perhaps the central issue from the point of view of cognitive science is, what happens when someone formulates a problem? We know that children invent their own strategies for solving problems; do they also invent strategies for formulating problems? Problem formulating appears to require facility in identifying the important features of a problem, abstracting from previous problems encountered, and seeing problems as organized into related classes. How much facility in each of these processes is required for problem formulating? What other processes are required?

Both problem formulating and problem solving appear to depend heavily on well-organized subject-matter knowledge. What is the relation between problem formulating, problem solving, and a structured knowledge base? How rich a knowledge base is needed for problem formulating? How much structure? How does experience in problem formulating add to the knowledge base?

To formulate a good problem and recognize it as good requires judgment and taste. How do people develop aesthetic standards for judging problems that they have not yet solved? How do they develop the ability to monitor and control their problem-formulating processes? What metacognitive processes are needed for problem formulating?

These questions illustrate the depth of our ignorance of problem-formulating abilities and how they are structured. To have stated the questions, however, is to have made a start toward answering them. As James Thurber once observed, "It is better to know some of the questions than all of the answers."

ACKNOWLEDGMENTS

I am grateful for comments on an earlier draft of this chapter from various conference participants, particularly Jim Greeno, Jim Kaput, Alan Schoenfeld, and Kurt VanLehn.

DISCUSSION

Kurt VanLehn. It seems to me that with the development of new mathematical tools, the professional use of mathematics can be changed so that mathematicians won't find themselves having to grind out answers to known questions. Not only professional mathematicians but physicists, engineers, computer scientists, and others who have to use mathematics for a specific purpose will find using programs like MACSIMA and muMATH much easier than using their textbooks to retrieve specific formulae. If that happens, much more of their effort will be spent formulating problems. So it's going to be critical for school systems to adjust instruction to this new usage of mathematics.

As always, there's a practical side to this issue and a purely mathematical side. Let me take the practical side—with the clothesline problem. Suppose I want to build a clothesline which I must fit into a 6-by-6-foot space because I've got to put it on my patio. What kind of design should I use? Issues like the quantity of wood, the structural stability, the hardware required, and the maximum amount of clothes it can hold then become important considerations. You've got to weigh all those issues. Problem formulating consists precisely in thinking of those questions ahead of time. Those questions lead to the realization that there are going to be knots, that the clothesline may be stable in one design and unstable in another, that one design spins and another doesn't. That kind of *preassociating* a problem is something that we don't know anything about yet.

Cognitive science hasn't addressed those questions, and for good reason. They are hard questions. But that might be a direction that the educational establishment will be forced to go in meeting the future demands of adults who want to use mathematics in an environment with sophisticated mathematical tools.

Jim Kaput.   There are problems, and there are problems. In problem formulation, I see two general classes that are somewhat overlapping. One class contains problems that arise when we are trying to model a situation. This kind of problem formulation, associated with the act of mathematical modelling, has not received much attention in the schools— certainly not the attention that it deserves. Then there is the other sort of problem formulation that uses the what-if-not strategy on an existing problem. An example here would be the Pythagorean triple idea discussed before. Those two different types of problem formulation situations each have their own attractions and rewards.

Let me make one other point regarding the technology. I think that the technology puts us in the position of being able to test our conjectures much faster than before because a machine will do some quick calculating for us and give us an output we can examine. We can let the parameters of the problem change in a systematic way, see what the resulting outputs are, and gain a higher level of mathematical understanding. That's a new possibility created by the use of the computer as a tool.

Alan Schoenfeld.   Following Jim, I have to note that there are various classes of problems but that students rarely work on real problems. Most of their work is on exercises that are called problems but that aren't problematic in the sense we've been discussing here. One issue that Jeremy alluded to briefly at the end of his paper was how you get problem formulation to actually take place in the schools. Certainly the reward structure that exists in the schools both for students and for teachers militates *against* paying attention to activities like problem formulation. For the most part, it militates against problem solving also. But there is even less of a reward structure for problem formulation.

In a class I sat in on last year, an exam was given in a 54-minute period that consisted of 25 locus and construction problems. The students had a total of 2 minutes and 10 seconds per problem. The teacher told the students before the exam that memorization is the way to handle these kinds of problems. The implicit message that students get is, "The way that you will do well in mathematics is to memorize the ways we solve problems." If that's the way you present problem solving, then how in the world are you going to get kids to formulate problems? There has to be some sort of reward structure both for teachers and students so they will receive some benefit from the activity.

Jim Greeno.   One point that Jeremy made should be underlined: We do have some reasons for thinking that successful problem formulation requires a great deal of knowledge in the problem domain. It also requires some different uses of that knowledge than are usually required for problem solving. I would suggest that, along with investigating flexible thinking of a general kind, we probably need to address questions of how people can acquire knowledge in a form that enables them to use it more flexibly. Flexible thinking probably isn't as content free as we have often tended to think. We usually think of giving children tasks about which they don't know anything and have to thrash around and be original. There probably is another process for being flexible and original with knowledge that you've spent a lot of time acquiring.

Jane Martin.   I have a hunch that the real problem formulators are primary grade students. Yet I think we have neglected this whole area in our research, and I would like to call for some research in problem solving and problem formulation at the primary level. I have noted that in kindergarten, first grade, and second grade, children will extend in an imaginative way the problems that are given. They are very open minded and very creative. I don't know what happens to them between then and junior high school, but I have found it more difficult to get junior high students to dream—either with a computer or with pencil and paper. I would urge us to look at the primary level for problem formulation. I would like to look at the possibility of meeting with teachers across the country, perhaps through a set of videotapes, so that teachers could receive some help in this area.

Roy Pea.   We don't know very much about how problem formulation activities go on in other areas such as science and report writing in social studies and history. How do children develop their ability to ask questions that are researchable? I know that little is known about this issue, but one success story is worth noting. The idea for the *Golden Books* emerged from someone's taking seriously children's questions about science. The authors studied the kinds of problems children formulated and wrote books that tried to address those problems. Maybe that is one kind of beginning.

REFERENCES

Anderson, J. R., Kline, P. J., & Beasley, C. M., Jr. (1979). A general learning theory and schema abstraction. In G. H. Bower (Ed.), *The psychology of learning and motivation* (Vol. 13, pp. 277–318). New York: Academic Press.

Arlin, P. K. (1977). Piagetian operations in problem finding. *Developmental Psychology, 13,* 297–298.

Branca, N. A. (1985). Mathematical problem solving: Lessons from the British experience. In E. A. Silver (Ed.), *Teaching and learning mathematical problem solving: Multiple research perspectives* (pp. 71–80). Hillsdale, NJ: Lawrence Erlbaum Associates.

Brown, S. I. (1984). The logic of problem generation: From morality and solving to deposing and rebellion. *For the Learning of Mathematics, 4*(1), 9–20.

Brown, S. I., & Walter, M. I. (1983). *The art of problem posing.* Philadelphia: Franklin Institute Press.

Bundy, A. (1983). *The computer modelling of mathematical reasoning.* London: Academic Press.

Burkhardt, H. (1981). *The real world and mathematics.* Glasgow: Blackie.

Caldwell, J. H. (1984). Syntax, content, and context variables in instruction. In G. A. Goldin & C. E. McClintock (Eds.), *Task variables in mathematical problem solving* (pp. 379–413). Philadelphia: Franklin Institute Press.

Calfee, R. (1981). Cognitive psychology and educational practice. *Review of Research in Education, 9,* 3–73.

Cohen, P. R., & Feigenbaum, E. A. (1982). *The handbook of artificial intelligence: Vol. III.* Los Altos, CA: William Kaufmann.

Cohen, S. A., & Stover, G. (1981). Effects of teaching sixth-grade students to modify format variables of math word problems. *Reading Research Quarterly, 16,* 175–200.

College Board (1985). *Academic preparation in mathematics: Teaching for transition from high school to college.* New York: Author.

Committee on the Undergraduate Program in Mathematics (1981). *Recommendations for a general mathematical sciences program.* Washington, DC: Mathematical Association of America.

Davis, P. J. (1985). What do I know? A study of mathematical self-awareness. *College Mathematics Journal, 16,* 22–41.

Davis, P. J., & Hersh, R. (1981). *The mathematical experience.* Boston: Birkhäuser.

De Corte, E., Vershaffel, L., & De Win, L. (1985). Influence of rewording verbal problems on children's problem representations and solutions. *Journal of Educational Psychology, 77,* 460–470.

Duncker, K. (1945). On problem-solving. *Psychological Monographs, 58*(5, Whole No. 270).

Frederiksen, N. (1984). Implications of cognitive theory for instruction in problem solving. *Review of Educational Research, 54,* 363–407.

Gentner, D. (1982). Are scientific analogies metaphors? In D. S. Miall (Ed.), *Metaphor: Problems and perspectives* (pp. 106–132). Atlantic Highlands, NJ: Humanities Press.

Getzels, J. W. (1964). Creative thinking, problem-solving, and instruction. In E. R. Hilgard (Ed.), *Theories of learning and instruction* (63rd Yearbook of the National Society for the Study of Education, Part 1, pp. 240–267). Chicago: University of Chicago Press.

Getzels, J. W. (1979). Problem finding: A theoretical note. *Cognitive Science, 3,* 167–172.

Getzels, J. W., & Csikszentmihalyi, M. (1976). *The creative vision: A longitudinal study of problem finding in art.* New York: Wiley.

Greeno, J. G. (1980). Trends in the theory of knowledge for problem solving. In D. T. Tuma & F. Reif (Eds.), *Problem solving and education: Issues in teaching and research* (pp. 9–23). Hillsdale, NJ: Lawrence Erlbaum Associates.

Greeno, J. G., Magone, M. E., & Chaiklin, S. (1979). Theory of constructions and set in problem solving. *Memory and Cognition, 7*, 445–461.

Jeffries, R., Turner, A. A., Polson, P. G., & Atwood, M. E. (1981). The processes involved in designing software. In J. R. Anderson (Ed.), *Cognitive skills and their acquisition* (pp. 255–283). Hillsdale, NJ: Lawrence Erlbaum Associates.

Keil, G. E. (1965). Writing and solving original problems as a means of improving verbal arithmetic problem solving ability. *Dissertation Abstracts, 25*, 7109–7110. (University Microfilms No. 65-2376)

Kilpatrick, J. (1985). Reflection and recursion. *Educational Studies in Mathematics, 16*, 1–26.

Kilpatrick, J., & Radatz, H. (1983). How teachers might make use of research on problem solving. *Zentralblatt für Didaktik der Mathematik, 15*, 151–155.

Krutetskii, V. A. (1976). *The psychology of mathematical abilities in schoolchildren.* Chicago: University of Chicago Press.

Lakatos, I. (1976). *Proofs and refutations: The logic of mathematical discovery.* Cambridge: Cambridge University Press.

Lenat, D. B. (1983). EURISKO: A program that learns new heuristics and domain concepts. *Artificial Intelligence, 21*, 61–98.

Lenat, D. B., & Brown, J. S. (1983, August). *Why AM and Eurisko appear to work.* Paper presented at AAAI-83 National Conference on Artificial Intelligence, Washington, DC.

Mason, J., Burton, L., & Stacey, K. (1982). *Thinking mathematically.* London: Addison-Wesley.

McCorduck, P. (1979). *Machines who think.* San Francisco: Freeman.

Miyake, N., & Norman, D. A. (1979). To ask a question, one must know enough to know what is not known. *Journal of Verbal Learning and Verbal Behavior, 18*, 357–364.

Novak, J. D., & Gowan, D. B. (1984). *Learning how to learn.* Cambridge: Cambridge University Press.

Perkins, D. N. (1981). *The mind's best work.* Cambridge, MA: Harvard University Press.

Pólya, G. (1954). *Mathematics and plausible reasoning* (2 vols.). Princeton: Princeton University Press.

Pólya, G. (1957). *How to solve it* (2nd ed.). New York: Doubleday.

Pólya, G. (1981). *Mathematical discovery* (combined ed.). New York: Wiley.

Resnick, L. B., & Ford, W. W. (1981). *The psychology of mathematics for instruction.* Hillsdale, NJ: Lawrence Erlbaum Associates.

Schoenfeld, A. H. (1985). *Mathematical problem solving.* Orlando, FL: Academic Press.

Schön, D. A. (1979). Generative metaphor: A perspective on problem-setting in social policy. In A. Ortony (Ed.), *Metaphor and thought* (pp. 254–283). Cambridge: Cambridge University Press.

School Mathematics Study Group (1970). Problem formulation. In *Secondary school mathematics: Unit 2. Teacher's commentary* (pp. 47–67). Stanford, CA: Author. (ERIC Document Reproduction Service No. ED 083 017)

Simon, H. A. (1973). The structure of ill structured problems. *Artificial Intelligence, 4*, 181–201.

Simon, H. A. (1983). Search and reasoning in problem solving. *Artificial Intelligence, 21*, 7–29.

Stover, G. B. (1982). Structural variables affecting mathematical word problem difficulty in sixth graders (Doctoral dissertation, University of San Francisco, 1979). *Dissertation Abstracts International, 42*, 5050A. (University Microfilms No. DA8211361)

Wilson, E. O. (1984). The drive to discovery. *American Scholar, 53*, 447–464.

6 From The Teacher's Side of the Desk

Anna Henderson
Maury High School
Norfolk, VA

Like all the chapters in this volume, this paper evolved from the discussions of a working group formed at our first conference. This working group, however (which was originally titled "The Teacher's Concerns"), was a little different from the others.

As Alan mentioned in the introduction, one of the main purposes of this project was to bring together members of two broad constituencies—researchers and practitioners—to communicate with each other. It has been very clear from the beginning of the project that the communication has to go both ways. One of the major themes behind this project was that cognitive science had developed a body of knowledge potentially useful to teachers, and that all of us working together could help to make that knowledge more usable and accessible. Another major theme was that, armed with an abundance of real-world experience, teachers could suggest some significant topics of inquiry for researchers. We teachers know all too well what kinds of situations cause difficulties for students and where help is desperately needed. A third theme was that teachers are important as a "reality check" for researchers. The context in a laboratory setting is very different from that of a classroom. For that reason, some kinds of behavior that occur frequently in the laboratory do not seem to be much of a factor in the classroom, and some classroom learning problems don't occur in laboratory situations. In short, close communication between the practitioners and researchers should be beneficial for both groups. Both groups have concerns that are uniquely their own, of course. But in the overlapping areas, the interplay between teachers and researchers should be useful for both.

For those of us who participated in this project, it has worked very much that way. A number of times over the past 6 months, for example, researchers discussing their work pointed to unusual behavior patterns they had seen in their experimental subjects. The behavior did not fit with the researchers' models, and explaining it was a problem. It turned out that the students' behavior made sense once their instructional history was known. If the students were taught the subject matter in a particular way, then they would be likely to try particular kinds of things when they generated the data for the researcher. (The following example gives the flavor of the discussions. Imagine trying to make sense of a student's algebraic scratchings on a piece of paper if the student was incorrectly using the FOIL method for multiplying two binomials, but you were not familiar with the method and did not know that it had been taught to the student the previous year.) In these kinds of discussions, teachers' descriptions of classroom or curricular practice explained some of the behavior patterns that researchers saw in the subjects' data. Cognitive scientists will agree that most laboratory studies take snapshot views of cognition, providing a view of what is going on in a child's head right now (or, at least, in the immediate past). Having a sense of the motion picture from which the snapshots are taken can be of tremendous assistance, and communicating the "larger picture" was one of the outcomes of our discussions.

Conversations about research findings worked in the other direction. In some cases, it was a matter of the research explaining, on its own terms, why students interpreted things the way they did in the classroom. In others, our discussions resulted in teachers' seeing the causes of what, until then, they had considered consistent but puzzling behavior. For example, it was during a discussion of subtraction bugs that Jane Martin realized a cause of some of her students' difficulties with fractions. It turned out that the incorrect step they took when subtracting one mixed fraction from another resulted from their having inappropriately generalized the "carry" procedure for subtracting whole numbers. (The details of this example are given in Steve Maurer's chapter.)

All of these issues, then, were among "the teacher's concerns." That such issues were the concerns of all of us was reflected in the fact that the group on teacher's concerns was a committee of the whole, whereas each of the other working groups for this conference had only six or seven members. But the issues just mentioned were far too many, and far too large, to handle in one conference session. To produce one paper, we narrowed the scope of our charge to something that seemed more reasonable. The working group on teacher concerns set out to list those areas that would most interest teachers and to delineate other matters of importance to teachers. We should stress that there is no topic from the school curriculum that cognitive researchers could work on in which teachers would not have an interest. There are, however, two or three general areas that need particular attention. Our concerns have to do with some areas that seem not to have been researched very much, and with questions about the best sequences and methods for teaching particular subject-matter topics. After discussing them, we will shift to another, larger issue that confronted us when we tried to establish our list of topics: the resources that are actually available to the teacher and the kinds of resources that will have to become available if we are to make the changes in classroom practice that will result in better learning for our students.

Let me begin with some topics of interest in the curriculum. When we asked our fellow teachers about what they thought was important, we discovered that we had so many lists of topics, and they were so long, that there was no subject-matter research topic for which there is not an audience. Nevertheless, the following comments point to a few general areas we feel need particular attention.

Material on early childhood learning through about fourth grade is readily available and helpful to the teacher. For instance, the topic of subtraction is a good example of where research has suggested a careful instructional sequence that teachers can use. The suggested teaching technique is a manipulative-based, hands-on concept development method that tries to help the student make connections between manipulations on some concrete materials (e.g. Dienes blocks) and the abstract paper-and-pencil algorithm. It works pretty well. Moreover, in subtraction there has been excellent work on bugs and on debugging strategies, which help teachers look for errors and be aware of aspects of their teaching that might lead to such errors.

This kind of work needs to be extended up into the upper elementary and middle school. Many concepts that are introduced in the fifth and sixth grade are not well researched. In those cases the teacher has no guide for classroom presentation except the rote algorithm. A case in point is division. True, there are arguments about whether long division should remain in the curriculum and that it may ultimately disappear, but in the meantime school systems expect teachers to teach the procedure. Those teachers have virtually no guidance on how to do so.

Another very serious problem that we face is how to teach the elementary arithmetic operations in upper elementary school, in middle school, and in high school to students who have never learned them. Again, some procedures, such as subtraction, have been well researched and are well documented; the sequence is there, and the approach is there, for students in the beginning grades. Should we use the same sequence and the same approach for a student who is in grade five or for a student who is in grade 10 and does not know how to subtract? This kind of problem occurs all the time and is not restricted to the K–12 curriculum. For example, a 1980 report from the Conference Board of Mathematical Sciences (CBMS, 1980) indicated that between 1975 and 1980 there had been a 400% increase in the number of remedial courses in mathematics at the college level. Predictions are that the 1985 CBMS report will describe an equally bleak picture. More and more young adults are entering college having mastered fewer and fewer of the mathematical skills they were supposed to have learned during their school years. Is more of the same instruction that led to failure in the first place going to help? Will something else help? We just don't know, and there is an urgent need to address such issues.

A related issue on which we could use help is on how to correct the mistakes that students make when they overgeneralize an operation. One example of this kind of behavior, probably the most frequent one, occurs with cross-multiplication. Students use cross-multiplication not only for comparing fractions and solving ratios, but also, inappropriately, for adding fractions, subtracting fractions, multiplying fractions, and dividing fractions. One of the things we need to know is whether the curriculum sequence is the culprit. Use of cross products is usually taught just before the students learn to multiply fractions. Does the proximity in the curriculum cause this kind of negative transfer, or is cross-multiplication simply so easy to do that it is applied nonselectively? Research that would help us to structure the curriculum and avoid negative transfer would be useful. But much more useful would be some sort of collaboration between researchers and curriculum developers that would produce well-designed curricula we could use. This brings me to the next major issue.

Reading the background papers for the conference and participating in the discussions was exciting for us teachers because the papers presented and the discussions echoed our own experiences and concerns in the classroom. The discussion of spiraling, for instance, suggested techniques that might be helpful in solving some classroom problems. The ideas are valuable, but turning those ideas into a classroom reality—doing the planning and development work necessary to incorporate those ideas into the curriculum—is another matter. The necessary planning and implementation take either preparation time or classroom time, and that is something most teachers simply do not have. It may take authority we do not have, or there may be other obstacles. Let me begin with time constraints.

After our first conference, I went back home and started to work on the list of important topics that would comprise this chapter. I asked the teachers in my school and in my district what they considered important. What topics would they like to have more information about, and what should we tell the researchers to work on? This lists that I compiled became incredibly long, as I mentioned earlier. But I saw something else too. To explain that, I need to say something about the audience for which we were researching—the teachers.

Most teachers want to teach well. They want students to learn. They want to provide the spark that motivates and the encouragement that supports. To paraphrase an earlier comment in our conversations, there are teachers and there are teachers. Of course, teachers have their limitations. Their attitudes vary from being willing to work for change and improvement, to wanting a quick fix, to not caring about a fix at all. Fortunately, the last group is not in very large supply.

But for all of these teachers, time is the enemy. Teachers have other responsibilities. Some must hold second jobs. Some are sponsors or coaches. Just everyday preparation and paperwork take an enormous amount of time. The demands on teachers create the conditions for their not wanting to experiment with something that may or may not be productive. It takes a great deal of extra time and energy to try something new in the classroom. If an idea the teacher tries does not work, then the whole class is further behind and the situation is that much worse.

I don't want to sound overly pessimistic. There are teachers out there who want to improve. There are supervisors who are supportive and helpful. I see teachers at conferences, taking night classes, and reading journals. All of them have an itch, and that itch is the desire to teach better. But when I spoke to my fellow teachers about the conference from which this book is an outgrowth, I heard the same thing from all of them: Even the best and the most dedicated of teachers is really strapped for time. The best may be the most strapped! Whatever time they manage to find during the school year is a blessing. It means they can be that much more careful grading papers, that they can give longer or more interesting assignments, and that maybe they will have the time to prepare some extra materials for a class. The everyday pressures and responsibilities of teaching do not give teachers the luxury to sit back and take lots of time to think about how to do things better. If they have any time, they will do the kinds of things that they know will get results, the kinds of things I just mentioned—spending more time on preparations and assignments. (See Sizer, 1985, for a more complete discussion.)

This volume is about communication between researchers and teachers. But the very best teachers have little time to go to the library to read research that is supposed to be most promising for school teachers. Even if they could find the articles that researchers think they ought to know about, they probably could not comprehend them, because most of those articles are written in jargon by researchers for researchers. The good teachers try hard. They go to NCTM meetings if they can get the time to go. And the teachers make the time to read the journals that are written for them. They read the *Mathematics Teacher* or the *Arithmetic Teacher.* They may read *Instructor,* and a small number may read the *Journal for Research in Mathematics Education.* That is all they have time for, period. If researchers don't reach the teachers through those journals or through workshops or their professional groups, they simply will not reach them.

Time isn't the teacher's only enemy. Sometimes there are circumstances beyond our control. We are in a changing society with a constantly shifting population. I teach in a Navy town. With some creative transfers we can get fourth or fifth grade students who have not had any work on fractions at all; they missed them in the various school systems into which they were transferred. I also teach in an inner city school with a transient population. Although there is a stable core of students all year, there is also another group who are on our side of town for three months and then are gone. And there are those who have moved into our area from the other side of town. So there is a constantly shifting population with everchanging deficiencies to be diagnosed and needs to be met.

As I mentioned earlier, changing how or what one teaches takes time; in some cases, it also takes authority that the teacher does not have. Teachers cannot arbitrarily change the sequence of instruction in their elementary classrooms, because they do not teach in isolation. They teach in conjunction with their students' previous teachers, the next teacher, and the other teachers in the building. In senior high school, we are a bit less tied down. If we want to try something new, we can cut out one topic and add another, for example, doing more problem solving and eliminating Cramer's rule. But the restrictions are nevertheless pretty tight. And some school systems have very rigid curricula that teachers must follow, with no discretion to do things differently. So it's possible for teachers to have wonderful ideas that they just cannot implement.

Then there are those school systems that practice benign neglect with little support. Teachers might be able to make some changes, but only on their own. There is much pressure to include technology, to include problem solving and estimation, various things covered in this volume, and many that are not. I am just now getting my feet wet on problem solving—and Jeremy Kilpatrick wants us to start on problem formulation! Roy Pea tells me it's important to deal with technology in the classroom, but I know teachers who are scared to death of a light switch, much less a computer. There are teachers in my building who have been teaching Algebra I for 25 years, and it is unlikely that they will ever have a computer in their classroom. And those teachers who want to try it will have to do the looking, the planning, and the incorporating on their own. The help does not come to them. They have to go looking for it.

Are we proposing a restructuring of instruction or a restructuring of the curriculum, or both? Certainly the first seems to be the more attainable goal. The political ramifications of wholesale curricular change are vast, and it is probably beyond our ability to effect that change. In either case, the main problem is to bring the research and the new ideas to teachers who may or may not be actively looking for help. The information needs to be readily available *and* accessible, and presented in plain language. Teachers are like students; they are unlikely to read any article that contains terminology they do not understand. They will not voluntarily add one more frustration to their lives. Without support from instructional leaders, change will not take place easily or evenly.

I know that the difficulty for the researcher is to write for the teacher so that the work is not only clear and readable but also retains the integrity of the research; we have seen how concerned Kurt VanLehn is, for example, that work on bugs not be misinterpreted or misapplied. I realize that it is a difficult assignment, but somehow we must find a way to make the connection between researchers and teachers. In some systems, supervisors and principals will help smooth the way; but many small school systems do not have people with the appropriate training. So teachers have to be the one audience for the work of researchers.

Another way to facilitate change is through textbooks. Major changes are highly unlikely, as we well know. The forces that shape publishing decisions are very restrictive. Authors to whom I have spoken are seldom able to write in complete harmony with their philosophy. There are some very good and very interesting books, but there does not seem to be much of a demand for them. Teachers' organizations need to take a stronger stand on what should be in textbooks. Researchers have to work with authors and publishers to influence textbook writing.

The one remaining liaison between teacher and researcher has to be teacher training programs, both preservice and inservice. Unless things have changed drastically since I was in one of those programs, most elementary teachers are not trained to be comfortable with mathematics, to know how children learn mathematics, and to make the best use of outside sources. Secondary teachers are very well versed in subject matter but not well trained in instructional methods. They don't know how to teach elementary processes except by rote, unless they have had good inservice training. This means that the researchers have to establish and maintain good connections with the people who are doing teacher training. The change that will come about from these kinds of connections will be slow, but it is important.

To sum things up: There are teachers out there who are aching to teach better. A comment Peggy Neal has made continues to haunt all of us. She said, *I know how to be a better teacher than I am.* For all teachers in that position, the best help from researchers would come from readily accessible and adaptable ideas and materials.

DISCUSSION

Jane Martin.   I'd like to return to the question of whether we should be concerned with restructuring instruction or restructuring curriculum. I would say that we have to do both, and I think we are in the midst of that. The notion of teaching approximation, estimation, problem solving, problem formulation, and of using new technologies in the elementary grades all imply a new curriculum. It's not just a revised curriculum; it's a new curriculum. Teachers are involved in trying to make changes all over the country, but there are some constraints. At lunch today I talked with a New York teacher who was concerned about the State Department of Education's taking a strong stand in writing objectives and actually dictating the direction of education in her state. There is a "Regents Curriculum" in New York, with statewide exams; and once the curriculum has been chosen, teachers and schools have very little discretion. I think that we, as math educators and researchers, need to examine how students learn and what they learn. We should not let the State Departments dictate to us. Instead, we have to dictate to them what needs to be done.

At the same time we must work to improve current instruction. Our teachers are not yet prepared to teach the current curriculum. We need a lot of help to be prepared to go into the classroom. For example, in modeling situations, we must learn to ask questions at the right time. Someone mentioned that there are questions to ask before, during, and after a student works a problem. But that view is *very* sophisticated. Some of the teachers I work with are not aware that they need to ask questions at all! My charge to preservice programs is to help young teachers coming into the profession understand that mathematics is not just a paper-and-pencil activity.

One instructional change I want to see is much greater use of students' verbalization. If students can explain to other students what they have done, if they can put some labels on their ideas and speak the language of mathematics, that is a sign that they can think mathematically. I believe that in our training programs we need to place more emphasis on talking mathematics, instead of on pushing a pencil on all those Ditto sheets.

Steve Maurer.   I came to the conference with the perspective of a teacher at the college or university level. During the conference I have been convinced that there is interesting new information to get out to teachers but that there is difficulty getting this information to them particularly at the collegiate level. College teachers labor under all the constraints that you mentioned, namely, that they are very busy and that what they can read is limited. But there is an additional constraint. College and university professors think they already know what they're doing. They think that unless somebody knows at least as much mathematics as they do, that person can't tell them anything about how to teach mathematics. How to disseminate research findings is a very difficult question.

Let me tell you a little bit about what I will read. Maybe I'm getting cranky in my old age, but I won't read anything unless the author makes it perfectly clear to me—in the first few paragraphs—that the author knows something that I don't know, about something that I want to know about. I won't read anything that's full of jargon. The terms that are used must be defined early in the paper in clear, simple language. And I won't read anything that is full of general abstractions and devoid of specific examples. The one exception to my rules is if I read a number of papers on the same topic, for example, in preparation for this conference.

There is also the problem that university faculty, like everybody else, tend to read certain journals or certain papers and not others. The mathematicians who read mostly professional journals may read about teaching only in the few pages of the *American Mathematical Monthly* devoted to the topic. That is, they will if they are MAA members. No AMS journal routinely deals with the topic. So someone has to tell us about this stuff, or we simply don't see it.

Now that I have made a number of negative points, let me make two positive remarks. One is that university and college faculty have more opportunities to attend professional meetings than school teachers. Appropriately offered talks at these meetings may be an effective way of communicating research ideas. Another is that university and college faculty can make changes much more easily than school teachers can. We have almost complete freedom with regard to instructional style, and—except in calculus classes in a big university where the same material is taught in 15 parallel sections—a lot of freedom with the curriculum. So if researchers manage to make some ideas interesting to us, we can try them out.

I mentioned earlier that I do feel that professional meetings are important forums in which research ideas can be communicated to people like me. I came with the additional skepticism that the general articles I had seen were telling me things I already knew, and which I assumed other faculty already knew. The conference discussions, however, have made me think about these issues in a new way. I can see applications of our discussions in my own classrooms, and I can see things that I want to change.

Let's take the topic of metacognition as an example. I remember at our first meeting my first reaction was, Please define what that is. It was a jargon term from my point of view. Once it was defined as "thinking about your thinking," I said, "Well, I ask my students to do that all the time." I enjoy teaching several different ways to attack problems and always ask my students at the beginning which method they think is the most effective way to do a particular problem. As we proceed through the method I ask them if we're going in the right direction. At the end I ask them if the answer is reasonable. So here I am patting myself on the back and noting all these wonderful things I do in the classroom. But now that I look back on my own teaching this semester, I realize that I've stopped doing something I used to do. I used to ask the class to suggest a method for doing the problem, which I would then pursue. A student once told me years back that he always had to pay attention because the fact that I was following a particular approach didn't mean it was the best one. I would continue with what they suggested until a point where I would ask if the approach still seemed fruitful. Well, this semester I didn't do that. I always started with what I knew was reasonable. I had cut out some of the ways I used to force students to ask if their approach was reasonable.

I thought about an exam I was writing. As a way of testing what students know, I often put on exams some questions that require the student to solve a problem three or four different ways. In reflecting on this conference, I realized that successful completion of this exercise shows that they know the methods, but it doesn't show that they considered the methods in advance and chose the most appropriate one for the problem. Perhaps I should write questions that require the student to solve the problem in the two ways they think will be best for the problem. Perhaps I should go a step further and grade them on their choice of method, rather than giving them points for whatever method they use, whether it's a wise choice or not. That's something I haven't done, and maybe I should.

In sum, I had thought about these issues and I thought I had a pretty good sense of teaching metacognitive approaches. Nonetheless, being forced to think of it again, I have seen gaps in my practice. I have seen places where I could improve, or at least try a different method. Being forced to confront these issues was good for me. It would also be good for a lot of my colleagues who haven't thought about them nearly as much. I urge that these issues be continually put before us. But they have to be presented in places where we look and in terminology that we can understand.

Warren Page. I want to raise some issues concerning the research dissemination and utilization paradigm that Steve takes for granted, in which researchers simply tell practitioners about the results of their research. I'm not sure it's healthy. If teachers are given the results of research, they become excited that they know how to become better teachers. But they can be thwarted in their efforts because of scarce resources or time, or political problems. If that happens, their expectations can turn against them and cause disenchantment or burnout. That is dangerous. In addition to describing research results, we need to work to provide teachers with the opportunity and the ability to include the research results in their practice.

More fundamentally, we need to rethink the role of teachers within the world of research and its dissemination. If teachers are viewed as partners in research, perhaps serving on teams with the researchers, then they make an investment in the research itself. That is far better than having them be passive consumers of research results. We need to find ways to include teachers as contributing partners in the research process. Everyone would gain from that change.

Peggy Neal. I want to return to an issue that Jane raised, that of preservice versus inservice training. Clearly, we can do better at preservice training than we have done. But most of our practicing teachers are finished with preservice, so you've got to approach the problem through inservice or staff development. It would take us a very long time to effect change if we went only through the preservice route.

I think there are unlimited possibilities for inservice development. My major reason for feeling that inservice is more valuable is my own experience. Nothing makes much sense until you become a teacher. Even to an eager student, even with some simulation, even with all the good things that you can do in teacher education, you just can't internalize what you're being told unless you've had a few years of real-life classroom experience. Inservice training is radically different. You can try some things in your own classroom immediately, and you have some experiential basis with which to appraise the ideas that are presented. I think we need a lot more inservice training.

Let me say some things about the attitude of classroom teachers. My own attitude toward the conference does, I think, reflect the attitude out there in the field. Andrea's description of our last meeting is a good start: I experienced depression, hostility, anxiety, frustration, changes of mind, and glimmers of hope. I would like to add one feeling from the classroom teachers' perspective—particularly since I represent them. That is one of excitement. I always find it very stimulating to be around people who are thinking, formulating problems, and searching for answers. For many reasons this is not the typical environment that I operate in. There are not many people around me that I perceive to be problem solvers or even good decision makers in any sense of the word. It's been a wonderful experience to have the opportunity to participate in a group like this.

However, the other emotions emerge too quickly and take control both in this conference and in others I go to. I become hostile when I hear the various condemnations of the educational system, and there have been quite a few at this conference. Then there is the depression when I realize that most of it's true. Then I feel frustration when I realize how little control I have over the entire educational system. There is anxiety about all of the new expectations that these ideas will engender—in fact, already engender—for teachers. I was thinking this morning about the new technology and all the new expectations of teachers that the computer brings. The political world and the intellectual world impose conflicting expectations on teachers. The long division algorithm and division of fractions are two examples. Everyone in this room probably agrees that these will eventually fall out of the curriculum and be replaced by something else. You expect me to be attuned to that development, and if I'm a well-informed teacher, I may be. But as Anna pointed out, my legislators in Georgia have not come near to accepting that change. It's not only the State Department of Education, but the members of the Legislature who have assumed a very active role in education. That brings us, then, to the changes the teachers at this conference are undergoing in our attitudes toward the field. We are ambivalent about them. It's a great feeling to experience them, but it's deadly to realize the increased expectations you now have to take home.

The last set of feelings is the glimmer of hope. My personal hope is that this conference, and others in the future, and the individuals who are here will, in fact, make a difference in my classroom as well as in classrooms all over the country. I have a cartoon that I began on the plane on the way up here but never completed, so I think I will describe it verbally. It has a teacher standing down here at the bottom of some steps. The teacher has two bags, one holding teaching techniques and the other, content knowledge. There is a crowd around the teacher, each person with different needs: supervisors, parents, administrators, and politicians. The teacher cannot even lift her bags up to open them. Her tricks are locked away, and the other people have got her surrounded.

Then there is a long escalator going up, and at the top are a number of doors with course numbers on them. There's Research on Cognition 101: Metacognition. There's Research on Cognition 102: Bugs. There's Research on Cognition 103: Representation. The person down here has her eyes on those doors and she is really excited. "Gee," she thinks, "if I could just get up there and find out what is behind those doors." I haven't filled in how that happens. I thought about drawing the next frame in which she pushes all those people out of the way, drops her little bags, and goes up there. Or she could just send a messenger up to knock on the door and say, "Send me something down." So I'm going to leave things at this point. As a teacher, I believe that there is immense potential in the educational system, because, as Anna said, most teachers do want to improve. They also have some respect for the part cognitive research can play in that whole endeavor. I hope, and I'm convinced, that the cartoon can be completed.

Alan Schoenfeld. Though it would be nice to leave "excitement" and "glimmers of hope" as our final emotional states, I am afraid there is at least one more emotion working on me right now. That emotion is fear. One thing that has become crystal clear to me during this conference is the enormity of the task we face. I want to talk about two issues, logistics and politics.

Taking logistics first, we are talking about effecting curricular and instructional change on a very large scale. Some of Henry Pollak's comments about the realities of implementing the major curriculum projects of the 1960s (e.g. the New Math) are relevant here. Roughly speaking, the figures working are these. Membership in the National Council of Teachers of Mathematics is approximately 60,000. Most NCTM members are secondary school mathematics teachers. All told, there are about 200,000 secondary school math teachers in the United States, so you have a good shot at getting information to them through NCTM or through the grapevine. With numbers like these, and a good support structure (for example, the large network of summer mathematics institutes sponsored by the National Science Foundation in the 1960s), effecting some meaningful change at the secondary level is a real possibility.

But when you turn to the elementary curriculum, the numbers work against you. Only a small fraction of NCTM members are elementary school teachers, perhaps 10,000 or so. Compare this number to the number of teachers at the elementary school level who teach mathematics: roughly two million. The teachers who are at this conference are the very, very best. The ones we see at professional meetings and who take enrichment courses—that is, the ones who are actively trying to get up the staircase Peggy described in her verbal cartoon, although they face tremendous obstacles—are still the very best, but a small minority at that. These people labor under all of the burdens that Anna described. Moreover, mathematics is just one of the many subjects for which elementary school teachers are responsible; they have limited time and resources to devote to any particular one of the subjects they teach. So we're faced with the fact that the dissemination mechanisms that would work at the secondary level don't work at the elementary level, and that teachers at the elementary level need more support and better prepared materials than their counterparts in the high schools. If we researchers are serious about wanting our work to have meaningful impact, we will have to deal with such issues.

Turning to politics, I want to say something about how curricula are designed and textbooks are written. Some recent experience of mine points to the difficulties we face. As Jane Martin noted, New York has a statewide curriculum for most academic subjects. The mathematics curriculum is now in transition, with the traditional sequence of courses (algebra in 9th grade, geometry in 10th, trig in 11th, etc.) being phased out and a new, "integrated" curriculum being phased in. Since problem solving is considered important these days, the people in charge of designing the curriculum decided it should be there.

Not long ago I got a telephone call from one of the two people responsible for revising the curricula for grades 7 and 8. It happened that we knew each other. We had been neighbors when I taught at Hamilton College, and had met when I gave a problem-solving workshop; I had come to her classes and we had worked together a little bit. When she was asked to work on the seventh and eighth grade curriculum, she decided to get in touch. We spent some time together, and I hope it was useful. I am fairly confident that the final product will be decent, because the people responsible for it are thoughtful and talented. But I am very concerned about the process that produced the curriculum. It turns out that I was the only researcher connected with the project in any way, and my involvement was purely accidental. There was no expectation of, and no provision for, any consultation with the research community in the design of the curriculum. The entire curriculum—to be used by virtually all students in New York State—was being designed as if the research community did not exist. Whether or not researchers had anything useful to say was a moot question; even if they did, it wouldn't have been heard anyway.

Along the same lines, it's worth repeating a comment about the locus of power in education. It's said that the most influential people on educational matters in the United States are the few dozen members of the textbook adoption committees in California, Texas, and New York. Since the markets in those states are tremendous, major publishers tailor their texts so that they're suitable to members of those committees. There is just one edition of each textbook series published nationwide, so what is acceptable to members of committees in those states becomes the norm for the country. If you have any illusions about the nature of the process that yields textbook adoptions, you might want to look at "Judging books by their covers," in Feynman (1985).

The point of these comments is that researchers cannot stand idly by on the sidelines, doing good work and expecting it to have an impact. Nor can teachers expect things to change for the better by themselves. There are tremendous logistical obstacles to effecting change, and the process by which the change will be made possible is a highly political one. If we want to see things happen right, teachers and researchers have to get involved in both the logistical and the political aspects of the situation.

Kurt VanLehn.   I was struck by Anna's saying she was satisfied with the research being done at the early primary level, because most of it looks at only two or three children doing interesting tasks in laboratory situations. Often the children are taken from the vicinity where the research is done, such as Berkeley, Pittsburgh, or Palo Alto. There's no middle layer of research between the teacher and cognitive research institutions. Do you trust that kind of research?

Anna Henderson.   Perhaps it's not just that kind of research we are seeing. All the articles are not necessarily that honest about where this research comes from. I may know that a single piece has limitations, but looking at the work of a number of people and combining it with what I already know gives me support for trying instructional innovation with particular groups of students.

Herb Ginsburg.   May I ask Kurt what a useful kind of middle-level cognitive research would look like?

Kurt VanLehn.   Well, I believe that such work is going on right now. I've reviewed grant proposals that call for that kind of work, so I suppose some of it is going on. For example, you might take an idea like metacognition and propose reifying the students' problem-solving strategies and doing drill and practice on choosing strategies. It sounds like a great idea, but I would want to see it tested with a wide variety of students across the country for 2 or 3 years. Then, if it seems that it actually helps, it could be implemented. By that time we would know much more about how to implement an instructional program, and we could make the teacher's job a little easier.

Herb Ginsburg.   You mean applied cognitive research, not cognitive research itself?

Kurt VanLehn.   Well, I guess so. In my industry, the software development industry, there is an analogous phenomenon. Designers develop wonderful ideas for programs that committees decide are saleable. Then the program outlines to into other people's hands and are reimplemented as production pieces of software. So my middle-level cognitive research would reimplement the cognitive researchers' ideas in a practical setting to find out how to make them work. I'm sure some of that goes on but probably not enough.

Joe Crosswhite.   I suspect that it goes on not because of cognitive science research but rather independently of it. The history of research on subtraction, for instance, is very old. I don't think it has been guided by those studies with one, two or three students at all. There is a body of research in mathematics education also. I will try to remind people of that in my reaction talk.

REFERENCES

Conference Board of the Mathematical Sciences. (1982). *Survey of undergraduate mathematics instruction*. Washington, DC: The Board.

Feynman, R. (1985). *Surely you're joking, Mr. Feynman*. New York: Norton.

Sizer, Theodore. (1985). *Horace's compromise: The dilemma of the American high school*. Boston, MA: Houghton Mifflin.

7

New Knowledge about Errors and New Views about Learners: What They Mean to Educators and More Educators Would Like to Know

Stephen B. Maurer
Department of Mathematics
Swarthmore College

One of the insights of the cognitive science approach to learning theory is that many of the mathematics errors students make are systematic. These errors are *bugs,* like bugs in computer programs, not slips.

So what? Astute teachers have known this for years (without the computer terminology). Also, astute researchers in traditional learning theory had documented this back in the 1920s.

The purpose of this paper is to answer that question—so what?—from the perspective of some mathematics educators who have had the opportunity to learn a bit about what the cognitive learning theorists are doing.

There *is* something new in today's statement that students make systematic errors. Briefly put (we say more later), researchers are now able to predict a large number of the arithmetic mistakes that individual students will make—before the students work the assigned problems! Hundreds of systematic error patterns have been identified. (On the other hand, more questions have been raised than answered. For instance, very little is known about the stability of these bugs; sometimes they are lasting, sometimes they self-correct and sometimes they reoccur spontaneously.)

In any event, it is not the specifics of this new knowledge which is really important to educators, not at least until the specifics are more complete. What *is* important is the general insight these studies give into how students learn. Indeed, "learn" may not be the best word to describe what happens; "interpret" may be better.

All told, the research brings Good News and Bad News. The Good News is that, basically, students are acting like creative young scientists, interpreting their lessons through their own generalizations. The Bad News is that their methods of generalizing are often faulty.

As a result of what we have learned, the educators among us have many additional questions we would like the learning theorists to address—about both the theory and its usefulness for education. These questions are typical results of the dialog which took place at the two conferences. This indicates that exchanges between learning theorists and educators can be beneficial to both groups.

Four sections follow. First we explain what's new. Then we highlight what's good and what's bad. In the third section we consider what all this means for the classroom, concluding with some recommendations. In the final section, we list questions for further research.

WHAT'S NEW

The cognitive science approach to learning involves the very detailed study of what *individuals* are thinking. This stands in sharp contrast to earlier educational work. Typical earlier research consisted of making statistical analyses, over large groups, of the effects of this or that educational activity—without much analysis of how and why the effects actually take place. Today there is much more detailed information than before about systematic student errors, and theory which says how the errors come about.

This "error analysis" is most complete for grade-school subtraction, to which many people-years of study have now been devoted (see Brown & Burton, 1978; Brown & VanLehn, 1982; & VanLehn, 1983). The attention to subtraction is not necessarily because subtraction has special importance, but in part because it is an appropriate domain for research—simple enough to be tractable but rich enough to allow for all sorts of wrongful behavior. Knowledge of the types of subtraction errors students commit, and why, is now so detailed that Artificial Intelligence computer programs have been written which make the same errors as students. (This is not accomplished by giving the computer detailed lists of what to do wrong; an AI program starts only with broad principles.) Other programs have been written which quickly diagnose which bugs a given student has. Still others help train people to diagnose other people's bugs. (It has proved useful, incidentally, to train students as well as teachers. In practicing their debugging skills, students come to recognize that their own reasoning can be buggy.) Also, the theory behind bug generation has begun to provide insight into better and worse choices of examples and better and worse ways to sequence the material. This was something the old knowledge of systematic errors could not do.

In any event, current research indicates that many answers to problems which have been considered slips (careless errors) are in fact the result of systematic misunderstandings about subtraction.

Figure 7.1 shows the work of a student with a certain bug. Clearly the student has occasional trouble with borrowing. Now, however, we can be much more precise.

This student has caught a version of the bug called "Borrow Across Zero". When s/he needs to borrow from 0, the student jumps across the 0 and borrows a 1 from the next (non-zero) column to the left. That is, given a subtraction of the form

$$\begin{array}{ccc} a & 0 & b \\ - \quad x & y & z \end{array} \quad \text{where z is larger than b,}$$

the student transforms the subtraction to

$$\begin{array}{ccc} (a-1) & 0 & (b+10) \\ - \quad x & y & z \end{array} \quad \text{instead of} \quad \begin{array}{ccc} (a-1) & 9 & (b+10) \\ - \quad x & y & z \end{array}$$

This procedure confronts the student with a second difficulty when it comes time to do the subtraction in the middle column. S/he then has the problem of subtracting y from 0, after all the borrowing s/he thinks necessary has already been done. In the particular version of the bug given in Figure 1 (which occurs frequently), the result of subtracting anything from 0 is 0. In another popular version, $0-6 = 6$ in the middle column. Students with this second version of the Borrow Across Zero bug would produce 117 instead of 107 as the answer to the third problem in Fig. 7.1, and 888 instead of 88 for the last problem. Otherwise they would get exactly the same answers as shown.

278	352	406	543	510	1023
− 135	− 146	− 219	− 367	− 238	− 835
143	206	107	176	272	88

FIG. 7.1

There is nothing random or sloppy about the work of students with bugs like these. The students are working carefully, following a precise— but incorrect—procedure. Test your understanding of students' bugs: what would a student with the first version of Borrow Across Zero obtain when subtracting 237 from 605? What about a student with the second version?

What accounts for such bugs? Roughly, the story goes like this. Imagine that the student understands the subtraction algorithm in a very mechanical way. S/he thinks about it as mere column by column manipulation, except that, if the top digit d of a column is greater than the bottom digit, one decrements the digit c to its left by one, and replaces d by $10 + d$. Imagine further that the teacher has never presented a case where c is 0, or that the student wasn't paying attention when the teacher did. What does this student do when s/he comes across an example with $c = 0$? S/he doesn't quit—s/he knows there should be an answer. So instead the student *generalizes,* figuring out some version of the procedure which includes all the previous cases and this case too. In this example the student's idea is that the digit borrowed from need not be *immediately* to the left, but may be the closest nonzero digit on the left.

This procedure is not just a random error but in a certain way makes sense. Most of the student's prior experience has probably been with 2-digit subtraction problems, in which carries were always taken *from the left.* Of course we know that the carries had to come from the adjacent column, because of the structure of base 10 arithmetic. But suppose the student doesn't understand that. If, instead, the student focuses on the procedure and tries to generalize it, the generalization may be: the carries come from *somewhere on the left.* In this way, confronted with a gap in the algorithm as s/he understands it (it doesn't say how to borrow from 0), the student incorrectly patches up and creates a bug. The study of such "corrections" is called *Repair Theory.* Since this theory is about *creating* bugs, not *eliminating* them, the name is unfortunate. *Gap* or *Patch* Theory would have been better.[1]

Continuing with examples, Fig. 7.2 illustrates another bug. Can you figure it out?

278	352	406	543	510	1023
− 135	− 146	− 219	− 367	− 238	− 835
143	214	213	224	328	1812

FIG. 7.2

This buggy procedure is: in each column subtract the smaller digit from the larger. This too, when looked at a certain way, is a natural generalization. Before the student was introduced to multidigit subtraction, s/he probably had many problems of the form "What is the difference between 7 and 3? . . . Between 4 and 9?" The rule for answering these questions was clearly to subtract the smaller from the larger. Furthermore, in the

---

[1] Repair Theory does not explain errors in the sort of language we have just used. The scenario imagined above—with students intentionally making generalizations along certain lines—should make the behavior correctly predicted by the theory both plausible and memorable. However, the theory itself makes no assumptions about intentionality and does not claim that the generalizations are consciously made. Rather, the theory is all couched in terms of various artificial intelligence constructs, and is quite technical. Interested readers should consult Brown and VanLehn (1982), VanLehn and Brown (1980) and VanLehn (1983)

first multidigit problems the student saw, the columns were probably handled independently, as in the first example in Fig. 7.2. All the column subtractions in those earlier multidigit problems were also, of course, bottom from top, but the student may not have seen that particular aspect as being significant. The teacher may unknowingly have abetted this view by thinking the bottom-from-top aspect so obvious as not to merit explicit mention that it must always be done this way. Under these circumstances, the bug "Smaller From Larger" is a natural generalization.

To date, scores of individual subtraction bugs have been identified. Thousands of grade school students have been tested, in the United States and overseas, and about 40% of students in grades 3 through 6 have consistent subtraction bugs (Brown & Burton, 1978, p 181). Moreover, a student can suffer from more than one bug at a time. In such cases, it would be difficult, without the benefit of bug theory, to look at such a student's paper and see any method in the madness. Try your hand now at Fig. 7.3. Then show it to someone else, without giving them the benefit of reading this paper!

278	352	406	543	510	1023
− 135	− 146	− 219	− 367	− 238	− 835
143	214	203	224	320	1012

FIG. 7.3

Bug-like errors have also been diagnosed in algebra, but not in as much detail (see Matz, 1982). Consider the student who does any of the "simplifications" in Fig. 7.4. This student has learned the distributive law all too well! We teachers emphasize over and over how we wish to simplify things by breaking them into parts and how the distributive law allows this; e.g.,

$$3(x+y) = 3x + 3y \quad \text{and} \quad (ab)^n = a^n b^n \quad \text{in high school algebra,}$$
$$L(x+y) = L(x) + L(y) \quad \text{in college linear algebra.}$$

The student who makes the unfortunate generalizations in Fig. 7.4 does not realize that distributivity works only some of the time.

$$(a + b)^2 = a^2 + b^2, \quad \frac{1}{a + b} = \frac{1}{a} + \frac{1}{b}$$
$$\sin (a + b) = \sin a + \sin b$$

FIG. 7.4

Another example of a systematic algebra error comes from the student's taking us literally when we say that letters just stand for numbers. A student, asked to evaluate XY when $X = 5$, $Y = -2$, comes up with 3, because substituting exactly, XY becomes $5-2$, which is 3. Or, a student who sees

$$\frac{aX + bY}{X + Y}$$

may simplify this to $a + b$ because that student has learned to replace the *pictograph*

$$\frac{X}{X}$$

by 1. The fact that X isn't being divided by X in the original fraction is not at issue to such a student. Only the pictograph counts; the meaning that lies behind the symbols is lost!

We said "bug-like errors" in algebra instead of "bugs" because it is not yet certain that there exist bugs in algebra in the precise sense. It is not clear that a student who, say, makes the first error in Fig. 7.4 will consistently make that error, the way a student with the Borrow Across Zero bug will consistently make that error. Algebra simplification skills are much more complicated than subtraction skills. It may be found, once algebra errors are studied enough from the bug perspective, that a student who makes the error $(a + b)^2 = a^2 + b^2$ does so only in certain contexts, but in those contexts always does it. Then that student does have a bug. In any event, it is already clear that such algebra errors as we have described are bug-like in that they arise from plausible modifications of standard procedures.

GOOD NEWS AND BAD NEWS

The Good News is what all this says about how students deal with the unknown. They try—imaginatively. Like scientists, they start with the cases they already know about. When they get stuck, they make plausible modifications which are intended to cover the new situations as well as the old.[2]

---

[2]The child as scientist (but not always a successful one) is a fundamental metaphor of cognitive science, not just for mathematics learning, but for all learning from the moment of birth. This metaphor is based on the insight that—unless we are all genetically "hard wired" to see things certain ways—in order to fathom the immense complexity of sense data that make up the world, children intuitively generate theories (about physical objects, about language, etc.) and test them out. This metaphor is basically positive, at least about human potential.

For instance, consider this passage from a paper by two psychologists studying how 4–9–year olds deal with problems about balancing blocks:

The construction of false theories or the overgeneralization of limited ones are in effect

The Bad News is the type of generalizations the students try. In all the buggy procedures observed to date the generalizations are based on "surface" properties rather than meaning. For instance, in Borrow Across Zero, the student seems to think of the digits only as marks on the page. S/he seems to say, "I'm supposed to decrement somewhere, and the column to the left won't work; the next column left of that should do just as well." Were the student thinking about what these symbols *mean*, i.e., that a 1 two columns over means ten times as much as a 1 one column over, the student would never propose generalizations like this, or at least would reject them quickly. In short, most student modifications are based on *syntactics*, the rules of symbol pushing, rather than *semantics*, the underlying meaning of the symbols.

Another piece of Bad News is that, unlike good scientists, most students don't consciously test their generalizations. They just use them. They may not even be aware that they *are* generalizing. The idea that rules have limited domains seems foreign to them.

Another way to think of the good and bad is this. Students don't learn, they interpret. They actively work to make sense of—find regularities and patterns in—the teacher's demonstrations and explanations. However, what students get out of a lesson may be quite different from what the teacher intends or even from what the teacher actually presents. The regularities the students pick up may be quite incidental to the intended lesson, e.g., apparent surface structure rather than meaning. Also, if the teacher has left a gap, for instance, s/he hasn't explained yet what to do if there is nothing to borrow from in the next column, the students will charge ahead and fill the gap. Consequently, if students get problems wrong, it may not be because they "haven't understood" some part of the procedure, but rather because they have already invented an alternative to that part.

Unfortunately, what students pick up from lessons is highly dependent on what presumptions they bring into the classroom. This point was not discussed above, but it's easy to see: For instance, if a student has the impression that mathematics consists solely of symbol pushing, then regardless of what the teacher says, the student may think about any new algorithm in only superficial ways. The teacher's attempts to "make sense" of the procedure will be seen as "mere justifications." Similarly, if

---

productive processes. Overgeneralization, a sometimes derogatory term, can be looked upon as the *creative simplification* of a problem by ignoring some of the complicating factors (such as weight in our study). This is implicit in the young child's behavior but could be intentional in the scientist's. Overgeneralization is not just a means to simplify but also to unify; it is then not surprising that the child and the scientist often refuse counterexamples since they complicate the unification process. (Karmiloff-Smith & Inhelder, 1974, quoted in Brown, Bransford, Ferrara, & Campione, 1983)

students believe that mathematics is understood only by geniuses and is supposed to be memorized by ordinary people like themselves, then they will memorize instead of trying to understand. Furthermore, students who have only been assigned problems that can be solved in a minute or two will eventually come to believe that all problems can be solved in just a few minutes—and will thus give up when confronted with a problem which takes appreciably more time (Schoenfeld, 1985).

The strong influence of prior experience is already well documented in physics. It turns out that even a college physics course often fails to dislodge the non-Newtonian naive mechanics that most people carry around with them (Larkin, McDermott, Simon & Simon, 1980; Mc-Closkey, 1983).

Continuing with News, perhaps the Best News for teachers is this: once made aware of the *fact* that students make systematic errors, and given a brief explanation about how such errors involve overgeneralization, even experienced teachers can have fresh insights which may help them teach. Several such insights were gained at our conferences. For instance, one of us saw an old error in handling mixed fractions in a new light. This teacher explained that she had often seen students do the sort of thing in Fig. 7.5, and wondered where it came from.

$$7\tfrac{3}{5} = 7\tfrac{24}{40} = 6\tfrac{124}{40}$$
$$-5\tfrac{7}{8} = -5\tfrac{35}{40} = -5\tfrac{35}{40}$$
$$\overline{\phantom{-5\tfrac{35}{40}}}$$
$$2\tfrac{89}{40}$$

FIG. 7.5

During our discussion she realized that this error was caused by the inappropriate use of a procedure from whole number subtraction. The students who produced the likes of Figure 7.5 had been taught to indicate borrowing in subtraction by writing in the 1, i.e.,

$$\begin{array}{r} 435 \\ -188 \end{array} \qquad \text{becomes} \qquad \begin{array}{r} 2 \\ 4\cancel{3}15 \\ -18\ 8 \end{array}$$

The fraction bug comes from assuming that inserting the symbol 1 is what borrowing is all about.

Unfortunately, the correct diagnosis of bugs is often hard to do. It is easy to jump too fast to wrong conclusions about which bugs a student has, just as students overgeneralize too fast. Both the good and the bad are very well stated in the original paper on bugs, Brown and Burton (1978):

> The importance of simply admitting that there may exist underlying bugs cannot be overstressed. Without appreciation of this fact, a teacher must view failure on a particular problem as either carelessness or total algorithm failure. In the first case, the predicated remediation is giving more problems, while in the second, it is going over the entire algorithm. When a student's bug (which may only manifest itself occasionally) is not recognized by the teacher, the teacher explains the errant behavior as carelessness, laziness, or worse, thereby often mistakenly lowering his opinions of the student's capabilities.
>
> From the student's viewpoint, the situation is much worse. He is following what he believes to be *the* correct algorithm and, seemingly at random, gets marked *wrong*. This situation can be exacerbated by improper diagnosis. For example, Johnnie subtracts 284 from 437 and gets 253 as an answer. "Of course", says the teacher "you forgot to subtract 1 from 4 in the hundreds place when you borrowed." Unfortunately Johnnie's algorithm is to subtract the smaller digit in each column from the larger. Johnnie does not have any idea what the teacher is talking about (he never "borrowed"!) and feels that he must be very stupid indeed not to understand. The teacher agrees with this assessment as none of his remediation has had any effect on Johnnie's performance. (pp. 167–168)

Despite the pessimistic sound of this extract, it suggests some more Good News: we can take a much more positive view of our unsuccessful students. Many of us might be inclined to think, as in the quote, that an errorful student is "just real dumb". But in fact, subtraction is not a unitary skill, and such a student may be just a bug away from having it all right. Furthermore, a kid can be smart and have bugs. We repeat: a bug is caused by an imaginative generalization; it's just imaginative in the wrong direction. Given this view, the teacher now has the interesting job of determining the bug and trying to fix it, rather than the dull job of simply repeating the right procedure to a dullard!

The bug explanation of errors does call into question our universal grading procedure: the more problems wrong, the lower the grade. This procedure is fine given the assumption that mistakes are caused by carelessness and forgetfulness. The more careless and forgetful students deserve lower grades, and sure enough they will miss more problems. Unfortunately, more bugs need not lead to more wrong answers—or one bug might cause a whole bunch. A better approach might be to construct tests with foreknowledge of the common bugs and aim for a good correlation between the number of wrong answers and the number of bugs. Another might be to grade students directly on how many bugs they have. But how is this number to be determined?

SIGNIFICANCE IN THE CLASSROOM: FIRST THOUGHTS

Classroom teachers would like to exterminate bugs. Better yet, they would like to keep them from ever entering the room! However, if there's one thing which is clear from recent research, it is that bugs can't be kept out simply by teaching how to do things right. Very well, but what should we teach in addition? We claim that this is not so obvious.

In this regard, it is instructive to report what happened when, at our first conference, one of us—an experienced teacher—was introduced to the concept of bugs through some role playing. Another of us—experienced with cognitive science research—pretended to be a buggy student. The teacher did what she would usually do in the classroom. When she saw that the student was having trouble with subtraction, she did not try to identify a bug and speak directly to it. Rather she asked questions to check whether the student understood the principles underlying subtraction and resorted to Dienes blocks in an attempt to get the student to see his errors and correct them. The student said he understood the subtraction manipulations. He also said he saw how the manipulations on Dienes blocks (exchanging a flat for 10 tens, etc.) corresponded to the carry operations in the subtraction algorithm. But when he performed the algorithm again, he made the same mistakes he'd made before. He said, "I see what you're saying, and I think I'm doing what I'm supposed to do, but then you tell me it's wrong." The two had reached an impasse. Showing the student how it all fit together was not enough; he heard what the teacher was saying, but because he interpreted it in his own way, the teacher's explanations were to no avail.

This session pointed out the limitations of what might be called the "traditional" and "sophisticated traditional" approaches to teaching the subject—at least after the student has run into difficulties. In the traditional view, if the student is making mistakes, it means the student hasn't gotten the procedure yet. The prescription is to give more practice. But if the student misinterprets explanations of the right procedure, seeing it demonstrated over again won't do any good. The sophisticated traditional approach doesn't stay at the procedural, symbol-manipulation level; it tries to go back to the root of the problem, the meaningfulness of the procedure. But this approach can meet the same roadblock (and, as research cited later indicates, the correspondence between Dienes blocks and the subtraction algorithm may not be as straightforward as we'd like).

Nonetheless, there is an important argument in favor of either traditional strategy. There simply isn't time for teachers to try to analyze and remediate on a one-by-one basis all the problems of all their students. Thus the question arises:

i) Is it a productive use of class time even to begin to analyze the bugs of some students?

An second argument for the sophisticated traditional strategy may also be put in the form of a question.

ii) Won't the bug go away, or never arise in the first place, if the relationship between the algorithm and the concepts are made meaningful and evident from the beginning?

As for the first question, many of us felt initially that the answer should be No; but after two conferences our view is somewhat different. No, it's not worth it for teachers to train to become expert bug diagnosticians; this takes a great deal of time and computer programs will do this pretty well anyway (as they already do for subtraction). But it *is* important for teachers to know that there *are* bugs and other systematic errors, for them to be familiar with the most common types and for them to be tuned in to looking for such things. Attaining this knowledge does not take long. Many teachers have already been introduced to subtraction bugs through an hour and a half of work with a bug simulation program and they found it very informative (Brown & Burton, 1978, p. 170). Such instruction will lead teachers to successfully analyze some of the bugs of their students.

More important, a little instruction will make teachers more conscious about how everything they do in the classroom relates to bugs; for instance, how the order in which they present topics and the examples they use can allow, and may even encourage, certain bugs to creep in. In short, in the absence of detailed cognitive science results about how to deal with bugs, our best hope is that good teachers will intuitively deal with the situation well, given that they are sensitized to it. (This might still be our best hope even when and if such detailed information is available!)

As for question ii), some recent research (Resnick, 1983, and Resnick & Omanson, in press) suggests that the answer may be an unpleasant No. Even when students received substantial special "mapping instruction" (in particular, work emphasizing the analogies between Dienes block subtraction and the usual paper and pencil algorithm), the gain in arithmetic skill over a control group was insignificant, except for a very short time after the special instruction. More precisely, the special group gained in ability to explain the concepts of subtraction, but this made no difference in their ability to avoid bugs in actually solving problems!

More experiments are being run, to see if other factors are involved. For instance, it may make a considerable difference whether mapping instruction is the original instruction or is used for secondary remediation. In any event, as Resnick (1983) writes, "the problem of linking conceptual understanding to procedural skill . . . seems likely to prove more complex than many of us once believed." Further research is slowly starting to unravel some of those complexities. For example, a very close look indicates that the subtraction algorithm is *not* a straightforward reflection of manipulations of Dienes blocks! (Schoenfeld, in press.)

This leave us in an awkward position, for simply treating bugs on their own mechanical level does not seem very promising either. If one bug is suppressed, in time another may pop up; indeed, research shows that students' bugs evolve even if no attempt is made to intervene (VanLehn, 1982). Clearly, the whole issue of how to remediate the misunderstandings underlying bugs needs much more work.

We summarize this section by making a few recommendations we felt confident about as a group.

A) Teachers should be made aware that bugs exist and be given a brief introduction to analyzing bugs. They should focus on what students are doing and why, not just on the answers they produce.

B) Text books and teachers' manuals should contain some information on error analysis. Ideally, they should also contain sample tests and exercise sets which cause the most common bugs to surface early.

C) Teachers should start to think about how the way in which they introduce sequences of topics might actually promote certain errors.

FURTHER RESEARCH EDUCATORS WOULD LIKE TO SEE

As part of the research-educator dialog, we educators present a few requests in the form of questions, with some discussion after each.

1)   Are there bugs in algebra?

We realize that algebra is a much harder domain for cognitive learning theorists to work in, but to the practitioner it is important to know how broadly the precise idea of bug extends. If there aren't real bugs in algebra, then only some of the messages from the study of buggy arithmetic pertain there. For instance, the idea that students overgeneralize extends, but the idea that students whose answers are way off might be wrong only in one small, clearly defined part of their procedures might not extend. We need to know which insights can be applied—not just to algebra, but to all parts of mathematics.

2)   Can resequencing material significantly reduce bugs?

Bug theory indicates that many bugs arise because students jump in and fill gaps which their teachers have not yet filled, or because a later topic looks superficially like an earlier one. Can different choices of examples, or new arrangements of topics, significantly reduce the opportunities for bugs to arise? Or will such changes merely make different bugs prevalent? If some sequences are better than others, are there general principles, explainable to educators on their own terms, which will allow the educators themselves to determine useful examples and effective topic-orders?

3)   How much should teachers worry about bugs anyway?

We seem to have announced an answer to this in recommendation A) above, but that was just an intuitive consensus statement about a minimum to be done. All the specific guidance we can get will be appreciated.

There are really several questions here. The first involves a broad skepticism, given the realities of the educational system, about how much difference additional knowledge from educational psychology can make. As one of us at this conference said, "I *know* how to be a better teacher than I am." In other words, is it worth making a significant effort to improve things through better teacher response to bug-like behavior when the result might be negligible compared to potential improvement through smaller class size, better discipline, more expertise on how to organize class time, more time for teachers to think, better teacher morale, etc.? In short,

3a)   How much can be gained in the classroom by heightened concern about bugs, given the current state of education?

Even assuming that attempts to improve student learning through improved learning theory are a good investment (instead of, or in addition to, attempting to improve the school environment), one must still question the attention to bugs in particular. Specifically, we ask

*3b)   How intrinsically serious are bugs? How long
lasting and stable are they?*

In other words, are bugs really dangerous wrong turns that require
prompt remediation? Do they result in long lasting misconceptions or
seriously interfere with subsequent learning? (There aren't good studies
of how much *adults* have bugs in arithmetic.) Or are bugs basically
harmless, disappearing over time as people come to understand what
arithmetic is all about. Going further, could bugs actually have a positive
side, comprising natural or even necessary steps in the normal learning
process? (Maybe the students we should be flunking are the ones who
don't have certain bugs—at least at certain stages!)

Even if bugs are a serious and lasting problem, is direct attention the
best way to help? What sort of attention? For all the time we teachers
spend on technique, we are really interested in understanding. As a
profession we have a fundamental belief that the way to ensure correct
learning of technique is to make students understand what they are doing.
Further research on how to remediate bugs may open a new door to
finding out whether this belief is justified.

In any event, in the belief that understanding is important, we ask the
following broader questions.

4)   How can we best get students to understand?

For instance, is it true that the more representations students have of
an operation (e.g., for subtraction Dienes blocks, the abacus, the Arabic
number system, the Roman number system), the better they will grasp the
underlying ideas? For slower students, if not all, it may be better to stick
to just one alternate representation.

*5)   Does better understanding result in better problem
performance?*

Teachers believe fundamentally that it does, but we have already
mentioned that there is some troubling evidence in this regard (Resnick,
1983). Students have more trouble making connections between different
representations than we once thought, or at least in using these connections to provide internal critics to catch wrong answers. How can we help
them make the connections and use the connections to improve their
performance?

There has already been some research on the nature of the connection.
For instance, VanLehn and Brown (1980) devised a measure of the

closeness of two representations of subtraction, and suggested that students might be taken through a sequence of representations, each pair differing in a single aspect which is thereby emphasized.

6) Why do students take such a mechanical approach to math?

Teachers have long complained that students take an unduly mechanical approach to mathematics. Perhaps today we are in a position to learn why. Is this approach inherent in the human psyche? Does it come about in early childhood as a means of adapting to the world at large? Or are we math teachers responsible? (There is, unfortunately, some evidence that this may be the case.)

It's interesting to note that the AI programs that mimic buggy students include internal "critics." That is, in attempting to generalize a procedure to a new case, the programs devise several generalizations, throwing them out if they do not meet certain constraints. Unfortunately, all constraints emphasized by the programs are syntactic. In subtraction, for example, they ask things like: is each column operated on?

Students have trouble developing internal semantic critics too. Perhaps the fact that cognitive scientists have trouble devising formal semantic critics says something about why students do. Perhaps if cognitive scientists could succeed at devising semantic critics, teachers would get some new ideas on how they could help students devise them. Indeed, the notion of internal critics is currently of great interest; see, for example, the various discussions of metacognition in this volume.

7) How can student groups be used to facilitate learning?

There is considerable current interest among teachers in having students work in small groups (Burns, 1981; Slavin, 1980). Having students work together appears to enhance the class environment and social interaction, and increases the learning that takes place. We have a hunch it could help with overcoming bugs or getting at the problems underneath. For instance, having students show each other what they are doing might help them to articulate their strategies. Teachers would thereby gain access to these strategies, and the students themselves would become aware that their strategies may differ from those of other students. The development of metacognition would be aided.

Unfortunately, there is little research on group interaction and almost none in the cognitive science tradition. How can groups be managed?

What is the role of the teacher in group interaction? What learning mechanisms take place in group situations that foster individual cognitive growth?

*8)   What is the importance of "process modeling" in
the classroom?*

By process modeling we mean the teacher's demonstration of the processes the students actually go through. The buggy studies have shown that simply having the teacher show how to do things right is not enough. Unless the meaning somehow gets across, process modeling provides only partial data, which the students interpret and generalize wrongly. Will it help to have the teacher be explicit about the things the students do that are leading them astray? There is no substantial research on this either.

Our final question asks nothing new, but it asks in a way which we hope will be inspirational to teachers and cognitive scientists alike. Recalling the basic Good News—students seek to generalize their experience—and the basic Bad News—they often base their generalizations on the wrong things—we ask:

9)   How can we enhance the scientist in the child?

READING

The references listed at the end of this chapter include an excellent introduction to cognitive learning theory (Resnick, 1984). Several other references are cognitive science papers on buggy algorithms. We don't recommend that the practicing teacher devote much precious time to reading the technical papers because they are largely concerned with issues which, as least for now, do not seem important in the classroom. For instance, since their goal is to describe a theory of bugs in such detail that a computer could generate them, they give considerable detail about AI matters. Also, much of the discussion in these papers concerns philosophy of science issues, that is, what has been proved about human behavior by the arguments in the paper, what can be proved by such arguments, what is merely suggested by them, etc. If at some time it is shown that these AI models mimic human learning behavior so carefully that one needs to understand them in detail in order to instruct students optimally, then the fine print will be necessary reading for teachers. Until then, only general knowledge about bugs, and about the learning approach which engenders them, appears to be necessary. In other words, the new detailed knowledge acquired to date by cognitive learning theorists is not (yet) very useful in the classroom. What *is* useful is the general knowledge (which is old, i.e., there are bugs), and the cognitive framework in which it now sits (which is new).

DISCUSSION

Kurt Van Lehn.   Research in third and fourth grades, and in remedial junior high classes where subtraction is being taught, has shown that roughly a third of student performance is correct, a third buggy, and a third unanalyzable. That's not just my research. It goes back to the results of Buswell and Johns, Brownell, and others. They found similar percentages when they were testing. I want to underscore that we have developed a technology for analyzing bugs automatically but that the conceptions were there already. The same is true for the question of where bugs come from. A lot of people have known for a long time that students have a tendency to overgeneralize syntactically things that they are taught. The theory of bug generation that the team at Xerox PARC is trying to develop is a finer grained analysis. First of all, the theory is an attempt to specify what, besides generalization, is happening that makes bugs occur. Second, in those cases where generalization does occur, it tries to explain what kinds occur and why. The theory, if successful, would be a more substantial contribution to our knowledge than simply knowing that generalization happens.

I am also concerned about the dissemination of the buggy research results. When research like this comes out, we try to make everybody aware of it. We assume it might suggest helpful techniques for classroom teaching, or it might develop more sensitivity in the teacher. I am holding out for a stronger view of the kind of help that should be provided by research. It would be more useful to have highly accurate theories of learning in which you can actually simulate a student's performance on a certain exercise or a certain curriculum over a period of time.

I would also like to see some research that accurately simulates the remediation process, so that we can answer the question of whether or not it helps to know which bug a student has developed. It may not help. We don't know. Currently, the tactic is to tell the educational community whatever we have found out about subtraction bugs. They may be able to use it right now, but the best way to use it is not clear. I hope that in the future we can come back with very precise answers about how bugs are generated and the best techniques for remediating them.

Andrea Petitto.   What I have to say is a reaction to the whole conference up to now rather than to this specific topic. Something Jim Greeno said in the previous panel discussion is relevant here. He said that learning arithmetic tends to extend the understanding of number from that of an element in a counting procedure to that of an entity that can be manipulated on its own, and that algebra extends the understanding of arithmetic from perceptions of single operations to that of relationships among arithmetic operations. What we've been hearing in this talk and in a number of others is that students tend to work very hard to make sense of procedures without connecting them to their underlying meaning. As Steve said, they are doing a lot of work on just the syntactic features of the procedures themselves. My conclusion from this is that we are missing something. I believe that Jim is correct when he says that as students learn arithmetic they develop their concept of number. But if what they are doing in learning arithmetic is just pushing little symbols around on paper and not connecting them to a number concept, the question arises of how the concept of number actually develops. Something is apparently happening, but we're not focusing on it in the research. Maybe teachers are getting at it in some way in the classroom, but it's not clear how the understanding develops, whether it's deliberately fostered, and, if so, how.

I suspect that one of the reasons we are missing some of these phenomena is that in both research and teaching we are focusing too much on solving problems, that is, on having students produce correct answers to problems or exercises via correct procedures. I don't want to say that solving problems is not important, because I think it's very important. But, I think we are locked up in a theoretical paradigm, and not all of the questions that we want to answer can be answered from inside. I don't have any good recommendations at this point, because I'm in there too. There are other theoretical paradigms, but I have a very hard time connecting them up with what's in this one. We are going to have to do some hard thinking about what it is that people do with their minds besides solve the problems that are presented to them on paper, in class, by teachers (or in other nice, clear-cut ways). I think something important is going on besides the phenomena that we have been focusing on and have been able to analyze so easily.

Roy Pea.   Our group began to realize just a short while ago the seriousness of our charge. The bugs literature seems to provide the most compelling data and therefore looks like a case in which basic research in cognitive science could have a positive impact on math education. What's interesting is the kind of dialogue that's gone on between educators and researchers concerning the ways in which this information can be used in practice. It would be instructive to learn from what has happened to the bugs research how information is transformed when it passes from research into practice or about how specific the information given to teachers needs to be. I also wonder if other findings from cognitive science, either present or future, will be as generalizable and useable as the bugs research. I worry about that.

I also wonder what other cognitive science findings will make as specific a contribution to instruction as the bugs research. There probably aren't many that can, but I'm willing to hear opinions to the contrary. Steve gave the bad news about the bugs research: He said that it is far too detailed. There are too many bugs, too many ways that they can combine, and so on. So the bugs must be modeled by computer. But this kind of research has more radical implications: it may provide new tools for teachers. If this buggy system ever gets to the commercial market place, it could radically improve our diagnoses of the misunderstandings that *individual students have,* in ways that teachers might be able to use. Teachers don't have to understand the mechanics of the system if its output is useful for their instructional purposes. Cognitive science research on mathematical understanding and its development could be useful in providing powerful tools that serve the diagnostic ends valued by teachers. We need to do some research on how that diagnostic information must be structured, and we have to do the research in areas besides subtraction. I would certainly like to see that happen, but it's not something that Steve seems to be recommending. It would be a different level at which cognitive science could have some impact on educational practice.

Steve Brown. I'd like to pursue that point. It's hard to know what cognitive science can contribute to an educator's understanding of errors. The fine tuning that Kurt spoke about might be helpful, but Herb raised some interesting questions about how much fine tuning is necessary for educators. The points that Steve made are helpful and important. I agree that students may make systematic errors, that the errors are syntactic rather than semantic, and that it may be a good idea for teachers to attend to these errors. But to invoke cognitive science for the purpose of reaching those conclusions may be like using a Mack truck to pull a kiddy car. My question is whether there are things in the field of cognitive science, other than either the conclusions or the tools, that could be helpful to educators. It is conceivable that educators could borrow some of the metaphors of cognitive science. I'm skeptical both about the conclusions and the use of tools.

Roy Pea. I think there are some methods of cognitive science that at least some teachers have discovered or used, although they are demanding of time. I think the "think aloud" protocol idea is extremely powerful.

Teachers have thought that students, for example, were understanding the computer programs they were writing until they had their students read through their programs and say what the computer was doing at each point. They recognized that much of what the students were producing were the trappings of knowledge, without comprehension. I think the protocol method can be useful for some purposes.

Jim Greeno. I suggest it is the results of cognitive science that are important to education. The point that Steve and Kurt were beginning to get to goes beyond describing these bugs. Kurt's work is an effort to find out how they get there. We knew little or nothing about that before. In fact, the problem hadn't even been raised. But it is something that we are beginning to understand, and it yields important insights into the way that learning works under the present circumstances. It seems to me that we can improve our current teaching practice in part by knowing how it is flawed.

Kurt Van Lehn. The learning theory described the origin of the bugs. The content of errors is related to the structure of the curriculum. If the curriculum were different, the bugs would be different. In particular, many bugs seem to be generated by an interaction between spiraling in the curriculum on one hand and diagnostic and placement testing on the other. A spiraling curriculum by itself is probably a good idea, and diagnostic testing by itself is probably a good idea. When you put the two of them together in the classroom, however, they interact in a way that produces bugs. This claim needs to be supported with more research, but that's the kind of precise result I'd like to see. It could cause some substantial changes in the way schools are run.

Jere Confrey. I want to point out something about errors that has not been mentioned. I think that everybody agrees that it is important and interesting to understand why some errors are made and how we can eliminate them. It would be interesting to see what else we can get from errors and what we can learn from them besides their causes. Identifying errors can provide the opportunity for doing something positive. Consider, for example, the errors with the distributive property that were mentioned today. Rather than just pointing them out to the student, we can use these errors as a starting point for reflecting with a student about distributive property. Why is it that some operations are distributive and others aren't? I have been doing some work with teachers recently, and errors have proved to be indeed good starting points for reflection about mathematical topics as well as about the nature of mathematics. It's also worth mentioning that the negative feelings about making errors might not arise if we stopped looking at errors simply as things to be eliminated.

Kurt Van Lehn.   In the previous talk, we heard a request for showing when you had to do an algebraic transformation, not just how to do the transformation. Every time students are shown when to do the transformation, it is in the context of a larger problem. They are seeing positive examples of when to do the transformation, but they are not being shown counterexamples. They can only guess when they shouldn't apply a transformation. Errors are exactly cases when transformations shouldn't be applied, so errors supply the discrimination instance that appears to be critical for learning the "when" part of transformation. The same thing is true for the other skills that we have been talking about. In a sense, errors can provide the route to more direct learning. Now, there is quite a bit of research on consolidation drill versus discrimination drill. Since we could say that errors provide discrimination drill in the context of a consolidation curriculum, errors might be used in interesting ways to make up for the deficiencies of an otherwise good approach.

Jane Martin.   I'd like to speak to the issue of exterminating bugs. While observing teachers I see a great deal of directed teaching and modeling of skills. I think teachers expect these approaches to be sufficient for student understanding. Students have been shown the meaning of the problem; they also know the process, so they should understand, and an assignment is given. I believe that what we need to do before giving the assignment is *to check* that students actually understand the model that we have presented; they may not. That change in itself should make a significant difference in the number of bugs that appear later on. The next stage of prevention would be to oversee the students' practice. I see far too much independent practice. Students are going home to practice skills that they have learned without any monitoring. That's not good enough. I believe we're really going to have to educate teachers and bug research is part of what they should learn about. They don't need to know all the bugs, but they do need to know that bugs exist. Most teachers probaly don't. Teachers at all levels need a lot more help understanding that they have to do more than explain, model, and give assignments.

Jere Confrey.   I have trouble with language that describes how people think in nonproblematic, empirical terms. One of my concerns about the conference so far has been the absence of discussion of cognitive theory. Roy's mention of Vygotsky was one little piece. The reference to constructivism today is another small piece of theory. When Kurt talked about the curriculum and the notions that he had about sequencing and spiraling, I gained a fair insight into how he is going about his work and what assumptions are driving it. I want to call for more discussion of the theoretical underpinnings of the work we do.

Roy Pea.   I have just a brief response. My own sense of why there is a lack of theoretical discussion has to do with the state of art and the first questions that one asks. The kind of work that's been reported has been primarily snapshot, frozen state, and nondevelopmental. Everyone will bristle at this, but I'd argue that what we have been doing is trying to understand what's happening at a given moment in time rather than how understanding develops in an individual over time. Those kinds of studies just haven't been done yet. Kurt has talked about some longitudinal studies he'd like to do that focus on how students develop particular bugs, what kinds of settings bugs develop in, and how the specific instructional acts of a teacher can remediate them. Now that kind of work would get at the kind of theoretical issues you are talking about. I don't know of existing work like that. Although, research on bugs may be further advanced than other research in cognitive science that has implications for mathematics education, it's still in the infant stages.

REFERENCES

Brown, A. L., Bransford, J. Ferrara, R., & Campione, J. (1983). Learning, remembering and understanding. In J. H. Flavell & E. M. Markem (Eds.), *Handbook of Child Psychology, Volume III* (pp. 77–166). New York: Wiley.

Brown, J. S., & Burton, R. R. (1978). Diagnostic models for procedural bugs in basic mathematical skills. *Cognitive Science, 2,* 155–192.

Brown, J. S., & VanLehn, K. (1982). Towards a generative theory of "bugs." In T. Carpenter, J. Moser, & T. Romberg (Eds.), *Addition and subtraction: A developmental perspective.* Hillsdale, NJ: Lawrence Erlbaum Associates.

Burns, M. (1981). Groups of four: Solving the management problem. *Learning* (September), 48–51.

Larkin, J. H., McDermott, J., Simon, D., & Simon, H. A. (1980). Expert and novice performance in solving physics problems. *Science, 208,* 1335–1342.

Matz, M. (1982). Towards a process model for high school algebra. In D. Sleeman & J. S. Brown (Eds.), *Intelligent tutoring systems.* New York: Academic Press.

McCloskey, M. (1983). Intuitive physics. *Scientific American, 248* 122–130.

Resnick, L. B. (1983). *Beyond error analysis: The role of understanding in elementary school arithmetic.* University of Pittsburgh, Learning Research and Development Center.

Resnick, L. B. (1984). Toward a cognitive theory of instruction. In S. Paris, G. Olsen, & H. Stevenson (Eds.), *Learning and motivation in the classroom.* Hillsdale, NJ: Lawrence Erlbaum Associates.

Resnick, L. B., & Omanson, S. F. (in press). Learning to understand arithmetic. In R.

Glaser (Ed.), *Advances in instructional psychology (Volume 3)*. Hillsdale, NJ: Lawrence Erlbaum Associates.

Schoenfeld, A. H. (1985). Metacognitive and epistemological issues in mathematical understanding. In E. A. Silver (Ed.), *Teaching and learning mathematical problem solving: Multiple research perspectives* (pp. 361–379). Hillsdale, NJ: Lawrence Erlbaum Associates.

Schoenfeld, A. H. (1986). On having and using geometric knowledge. In J. Hiebert (Ed.), *Conceptual and procedural knowledge: The case of mathematics*. Hillsdale, NJ: Lawrence Erlbaum Associates.

Slavin, R. E. (1980). *Using student team learning* (Rev. Ed.). Johns Hopkins University, Center for Social Organization of Schools.

VanLehn, K. (1982). Bugs are not enough: Empirical studies of bugs, impasses and repairs in procedural skills. *Journal of Mathematical Behavior, 3*, No. 2.

VanLehn, K. (1983). On the representation of procedures in repair theory. In H. P. Ginsburg (Ed.), *The development of mathematics thinking* (pp. 197–252). New York: Academic Press.

VanLehn, K., & Brown, J. S. (1980). Planning nets: A representation for formalizing analogies and semantic models of procedural skills. In R. E. Snow, P. A. Federico, and W. E. Montague (Eds.), *Aptitude, learning and instruction*, Volume 2, *Cognitive process analyses of learning and problem solving* (pp. 95–137). Hillsdale, NJ: Lawrence Erlbaum Associates.

8 What's All the Fuss About Metacognition?

Alan H. Schoenfeld
Education and Mathematics
The University of California, Berkeley

This chapter was written after the second of our two conferences, in response to a challenge from (among others) Joe Crosswhite, Henry Pollak, Anna Henderson, and Steve Maurer. You'll find their comments in various places throughout this book. Their challenge can be summarized as follows:

> Metacognition is a buzzword for you researchers. Over the past few days [of the conference] we've heard about metacognitive this, metacognitive that, metacognitive the other. The word has been used in almost every talk, and almost every panel discussion, since we got started. But the plain fact is that it's jargon and doesn't communicate anything to us nonresearchers. If metacognition is so important, you have a responsibility to explain to us (a) what it is, (b) why it's important, and (c) what to do about it—all in clear language that we can understand.

What follows is an attempt to do just that. The next section discusses (a) and (b) together, defining metacognition and explaining why it is worth worrying about. The section that follows deals with (c), describing some of the classroom techniques I use to help students develop the "right" metacognitive skills.

WHAT METACOGNITION IS AND WHY IT'S IMPORTANT

Translating the term "metacognition" into everyday language, one gets something like "reflections on cognition" or "thinking about your own thinking." Although those definitions are in the ballpark, they're not precise enough to be useful. More precisely, research on metacognition has focused on three related but distinct categories of intellectual behavior:

1. Your knowledge about your own thought processes. How accurate are you in describing your own thinking?

2. Control, or self-regulation. How well do you keep track of what you're doing when (for example) you're solving problems, and how well (if at all) do you use the input from those observations to guide your problem solving actions?

3. Beliefs and intuitions. What ideas about mathematics do you bring to your work in mathematics, and how does that shape the way that you do mathematics?

There is a large body of fascinating research on the first item. Because most of the work with direct implications for mathematics educators has been focused on the second and third categories, however, we are going to skim briefly over the first. To sum it up in a few words, the research indicates that children are not very good at describing their own mental abilities, but that they get better (though nowhere near perfect) as they get older. The largest body of research is on "metamemory," or people's ability to describe how good they are at remembering things. Young children have very little idea of how well they can memorize. Although they may say (and believe) that they can easily memorize a hundred unrelated words, in fact they will have trouble memorizing more than four or five. As children get older, their estimates of their memory skills become more and more accurate.

Why is work in this area so important? For one thing, we're interested in helping students develop good study skills. Those skills depend, in part, on students' ability to make realistic assessments of what they can learn. Hence, it is important to know how likely it is that students will reflect on their thinking and how accurate those reflections will be. Similarly, good problem solving calls for using efficiently what you know; if you don't have a good sense of what you know, you may find it difficult to be an efficient problem solver. (I shall have no more to say about this aspect of metacognition in this chapter. For those who wish to pursue it, two good entry points into the literature are articles by Brown, 1978, and Brown, Bransford, Ferrara, and Campione, 1983.)

We turn to the second aspect of metacognition, control, or self-regulation. Another way to look at this aspect of metacognition is to think of it as a management issue: How well do you manage your time and effort as you are working on complex tasks? Aspects of management include (a) making sure that you understand what a problem is all about before you hastily attempt a solution; (b) planning; (c) monitoring, or keeping track of how well things are going during a solution; and (d) allocating resources, or deciding what to do, and for how long, as you work on the problem. Let me begin with two brief introductory examples and then broaden the discussion. The first example comes from calculus, the second from algebra.

A number of years ago I prepared an examination on techniques of integration for a calculus class. The exam began with the following problem:

$$\int \frac{x}{x^2 - 9} dx$$

I had chosen this problem quite carefully, expecting it to be a "confidence builder" for the students. If you note the relationship between numerator and denominator (the numerator is $\frac{1}{2}$ the derivative of the denominator), you can rewrite this problem as

$$\frac{1}{2} \int \frac{2x}{x^2 - 9} dx,$$

and solve it in one step using the substitution $u = (x^2 - 9)$. Substitution being the most basic technique of integration, the students certainly knew how to use it. I expected them to solve the problem in 2 minutes or less, feel good about it, and move on with confidence to the rest of the exam.

The exam was given to a large lecture class. About half the students did what I had hoped. But nearly one fourth of the students solved the problem using a technique called "partial fractions," in which they had to do some fairly complicated algebra to re-express the fraction $x/(x^2 - 9)$ in the form $[A/(x - 3) + B/(x + 3)]$. This technique works, but it takes most students at least 5 and perhaps 10 minutes to carry it out. But the time these students finished the first problem, they were behind schedule and scrambling to finish the test. Even worse, more than one ninth of the students solved the problem by using the trigonometric substitution $x = 3\sin\Theta$. This substitution also works—but it takes most students 10–15 minutes to use it. The students who did so wound up very far behind and did quite poorly on the test.

The point of these examples is not that the students failed to understand the mathematics. If anything, the students who used partial fractions and trig substitutions demonstrated mastery of more difficult subject matter than did the ones who used the simple substitution. The issue here is not what they knew—it's what they decided to use. The students who used partial fractions and trig substitutions violated a cardinal rule of problem solving: *Never use any difficult techniques before checking to see whether simple techniques will do the job.* Had the students asked themselves if there might be an easy way to do the problem, they might have avoided a great deal of unnecessary work.

I shall not give all the gory details of my algebra example; a brief description is enough for all mathematics teachers to recognize the behavior it discusses. In a typical videotaping session, some students were asked to solve the algebraic equation

$$x^2y + 2xy + x - y = 2$$

for y. They began by moving the lone y over to the right-hand side of the equation, which seems reasonable. They brought together the y terms on the left-hand side of the equation, factored out the common y, and then factored the common x term from that expression as well. Then they moved some terms over to the right-hand side . . . and so on. Each manipulation was correct. But a close look at what they did reveals that the equation they were working on after six algegraic steps was more complicated than the one they had started with! They simply proceeded one step at a time. Each time they got a new equation they seemed to start over: "Now what can we do with *this* equation?" As a result, they dug themselves into a deeper and deeper hole, without standing back to see if what they were doing made sense. This kind of behavior occurs all too frequently. (For a more extended treatment of students' algebraic foibles, see Ron Wenger's discussion in chapter 9.)

Both of these examples illustrate the main point about self-regulation: It's not only what you know, but how you use it (if at all) that matters. The discussion of Figs. 8.1 and 8.2 provide dramatic illustrations of this point. These discussions provide summary outlines of two representative problem sessions. Readers who would like all the details can find them in Chapters 4 and 9 of my (1985) *Mathematical Problem Solving.*

In my problem-solving research I frequently make videotapes of students solving problems, and then analyze their solutions in detail. Figure 8.1 presents the graph of an attempt by two students to solve a reasonably difficult problem: finding the largest triangle that can be inscribed in a circle. As it happens, this is a standard "max–min" problem, just like one that one of the students had solved correctly on a calculus final exam the week before. (He got an A in the course.) Both students knew more than enough mathematics to solve the problem without difficulty. But this problem was given out of context, and that made a big difference.

In brief, this is what happened. The students read the problem, made a conjecture (the correct one, that the largest triangle is equilateral), and then set out to calculate the area of the equilateral triangle. They made some mistakes and got bogged down in the calculations. They were still enmeshed in those calculations 20 minutes later, when the video equipment recording them clicked off. At that point, I told them the area of the

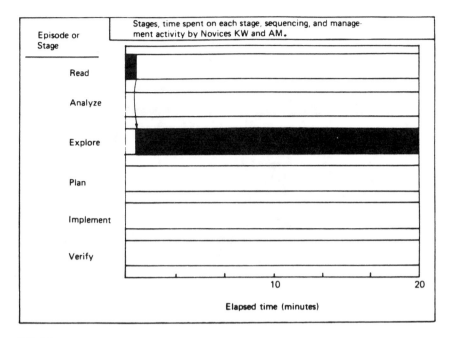

FIG. 8.1.

triangle and asked how knowing it would help them solve the original problem. They couldn't tell me.

The students had spent twenty minutes on a wild goose chase. They had ample opportunity to stop during that time and ask themselves "Is this getting us anywhere? Should we try something else?" but they didn't. And as long as they didn't, they were guaranteed not to solve the problem.

This is an all too typical example of the disastrous consequences of an absence of self-regulation. In an extensive series of videotaping sessions where students were recorded solving nonstandard problems outside the classroom, I found that more than half the problem sessions were of the type illustrated in Fig. 8.1, where students read the problem, decided to do something, and then kept at it to the exclusion of all other possibilities. The students were remarkably consistent—and persistent—in pursuing wild mathematical geese. Because I was familiar with their backgrounds, I had evidence that they knew the mathematics that would provide solutions to the problems they worked. Because of their wild goose chases, however, they never got around to using it. The result was an extremely high failure rate.

Figure 8.2, which shows the graph of an attempt by a mathematician to solve a difficult geometry problem, provides an interesting contrast. The

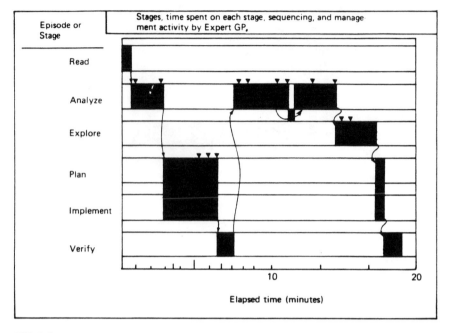

FIG. 8.2.

mathematician had not worked in geometry for a number of years, and he was pretty rusty when he started working the problem. He was far from a geometry expert, but he was an expert at problem solving. Interesting about his solution is that as he worked his way through the problem, he considered quite a few different approaches to it, many of which were just plain wrong. In fact, he generated enough potential wild geese to keep dozens of problem solvers busy. Yet he never went off the deep end as the students had, because he was as ruthless about testing and rejecting ideas as he was ingenious in generating them.

The problem that the mathematician worked had two parts, which explains the bipartite nature of the graph in Fig. 8.2. One difference (besides the fact that he solved the problem) between this graph and the graph of the students' work is immediately apparent: The mathematician spent the vast majority of his time *thinking* rather than *doing*. More precisely, he spent a great deal of time analyzing the problem. When he had made sense of it, working out the details of the solution didn't take long at all. The second feature distinguishing between the graphs is that there are a large number of small inverted triangles in Fig. 8.2, and none at all in Fig. 8.1. Each of those triangles indicates a time when the problem solver effectively pulled up short, asked "How am I doing?" and decided what to do as a result. The two marks at the beginning and end of the first "analyze" phase, for example, correspond respectively to comments about the need to make sense of the problem and the fact that (after analysis) he understood it fairly well and was ready to try out a solution. The comments near the end of the first plan/implement phase had to do with his being near a solution and the need to check it. He did solve the problem, and he checked his solution. When he began working on the second part of the problem he commented that it looked tricky and that he'd better be careful. That turned out to be true. He started off on a wild goose chase, but—and this is absolutely critical—he curtailed it quickly ("But I don't like that. It doesn't seem the way to go."), changed direction, and went on to find a solution.

In sum, the difference between the mathematician's success and the students' failure cannot be attributed to a difference in knowledge of subject matter. Indeed, the students started off with a clear advantage over the mathematician. They knew all of the procedures required to solve the problem they were given, whereas he did not remember them and had to figure them out for himself. What made the difference was how the problem solvers made use of what they did know. The students decided to try something and went off on a wild goose chase, never to return. The mathematician tried many approaches, but only briefly if they didn't seem to work. With the efficient use of self-monitoring and self-regulation, he solved a problem that many students—who knew a lot more geometry than he did—failed to solve.

We now turn to the third aspect of metacognition, beliefs and intuitions. These ideas were introduced in our discussion of constructivism in chapter 1. To recapitulate, people are *interpreters* of the world around them. They don't necessarily see "what's out there"—some version of "objective reality"—but instead perceive what they experience in the light of interpretive frameworks they have developed. As we said in chapter 1, this has significant implications for our teaching. When we try to teach some new subject matter, we can't assume that our students are empty containers waiting to be filled with knowledge. The students may have preconceptions and misconceptions about much of the subject matter they study, and we would do well to take that into account. Moreover, as the literature on bugs indicates, students may consistently misinterpret the procedures they learn in our classrooms.

The kinds of beliefs discussed here are more subtle, and more distressing. I argue that, largely as a result of their instruction, many students develop some beliefs about "what mathematics is all about" that are just plain wrong—and that those beliefs have a very strong negative effect on the students' mathematical behavior. Two examples illustrate the point.

The first example comes from the third National Assessment of Educational Progress (Carpenter, Lindquist, Matthews, & Silver, 1983). One of the problems on the NAEP secondary mathematics exam, which was administered to a stratified sample of 45,000 students nationwide, was the following: An army bus holds 36 soldiers. If 1128 soldiers are being bused to their training site. How many buses are needed?

Seventy percent of the students who took the exam set up the correct long division and performed it correctly. However, the following are the answers those students gave to the question of *how many buses are needed*: 29% said the number of buses needed is "31 remainder 12"; 18% said the number of buses needed is "31"; 23% said the number of buses needed is "32," which is correct; (30% did not do the computation correctly).

It's frightening enough that fewer than one fourth of the students got the right answer. More frightening is that almost one out of three students said that the number of buses needed is "31 remainder 12." Those students picked some numbers and an operation from the problem, did the computation, and wrote down the answer—without checking (or at least, without checking carefully) to see if the result made sense! In essence, they treated the problem as calling for a formal computation. Despite the "cover story" about the buses, the computation had little or nothing to do with the real world.

The second example comes from my research in geometry. I have given students a series of geometry problems, in which they were asked (a) to prove that certain geometrical figures have certain properties, and (b) to construct those figures. In a proof problem, for example, the students show that the center of a given circle lies at the intersection of two particular line segments. The students are then given the same diagram without the circle and are asked how to construct the circle. Of course, this second problem is not a problem at all: the proof states what properties the circle must have and therefore how to construct it. Yet about 30% of the students who solve the proof problem correctly then make a conjecture that flatly violates what they have just proven (Fig. 8.3). In other words, they ignore the results of the proof when working the construction problem. Why? These students see little or no connection between the two problems. From their point of view, proof problems confirm what they already know or what they have been told is true. Construction problems ask you to find something. Thus, when they work construction problems and are in "discovery mode," the results of proof—or "confirmation mode"—are simply irrelevant.

Much of my research in recent years has been devoted to exploring students' beliefs about mathematics, and I have spent a great deal of time making classroom observations to find the origins of those beliefs. Details may be found in my (1985) *Mathematical Problem Solving* and in my (in press) article, "When good teaching leads to bad results." Some of the unpleasant findings are as follows.

Despite having proved -- using the formal tools of Euclidean Geometry -- that the center of the circle tangent to two given lines lies at the intersection of the two perpendiculars and the angle bisector indicated in the following diagram,

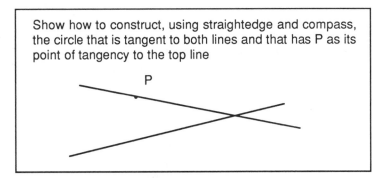

Thirty percent of the students asked to solve this problem:

Show how to construct, using straightedge and compass, the circle that is tangent to both lines and that has P as its point of tangency to the top line

P

conjectured that the center of the circle was the midpoint of the line segment that joined P and its "opposite," P'.

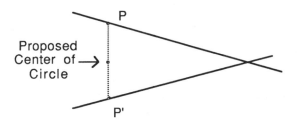

Proposed
Center of ⟶
Circle

P

P'

FIG. 8.3.

Many students come to believe that school mathematics consists of mastering formal procedures that are completely divorced from real life, from discovery, and from problem solving. This is the kind of belief behind the behavior in the two examples just discussed. After all, if mathematics problems were meaningful to students, could students possibly report that the number of buses they need is 31-remainder-12? (A colleague has pointed out that a group of students in the schoolyard, deciding how many cars they needed to go someplace, would never make the same mistake.) Similarly, students disregard proof because it's meaningless to them. Some other student beliefs—stated harshly to be provocative—are as follows: All problems can be solved in 10 minutes or less; the form of a mathematical argument is more important than its correctness; only geniuses are capable of discovering mathematics.

These beliefs, like the one discussed earlier, have unfortunate behavioral consequences. Students who believe that all problems can be solved in ten minutes or less will simply stop working on a problem after a few minutes, even if they would have been able to solve it with more effort. Students concerned with form (e.g., writing two-column proofs in geometry with correct abbreviations such as "CPCT") will spend more time worrying about the form of their answer than they will trying to understand the result they are writing down. And students who believe that mathematical understanding is simply beyond ordinary mortals like themselves become passive consumers of mathematics, accepting and memorizing what is handed to them without attempting to make sense of it on their own.

These example show that beliefs and intuitions, like self-awareness and self-regulation, are important determinants of students' mathematical behavior. "Knowing" a lot of mathematics may not do students much good if their beliefs keep them from using it. Moreover, students who lack good self-regulation skills still may go off on wild goose chases and never have the opportunity to exploit what they have learned. In sum, metacognition deserves our attention. Now, what do we do about it?

A "KITCHEN SINK" APPROACH TO DEVELOPING METACOGNITIVE SKILLS

This section describes four classroom techniques that focus on metacognition, and the rationales for them. The techniques were developed in my problem-solving courses, and (since I have a lot of flexibility there) it's easiest to use them in that setting. However, all of them can be adapted for use in virtually any mathematics instruction. We start with the technique that is least "interventionist" and proceed to techniques that call for increasingly deeper interactions between the teacher and students.

First Technique: Using Videotapes

Students asked what they do when they work on problems typically will say that they do "what comes to mind"—as though their minds are independent, autonomous entities over which they have little or no control. Of course this last statement is an exaggeration, but only a small one. The fact is that most students are largely unaware of their thinking processes. Virtually none of the students who enter my problem-solving classes are aware that they can practice their thinking skills and get better at them. Yet self-awareness is a crucial aspect of metacognition, for awareness of one's intellectual behavior is a prerequisite for working to change it. For that reason I make it a point to bring the subject out in the open.

Early in the term I describe the kinds of things we'll be doing during the term and why we'll be doing them. This introduction to the course includes a discussion of problem-solving strategies and of issues such as self-regulation and belief. My job, in part, is to convince the students of the importance of these aspects of mathematical thinking.

It's easy to make a good case for the problem-solving strategies to be studied in the course. The first day of class I give the class some carefully chosen problems to work on. I know from experience that the students could work on these problems for hours without success—but that the problems can be "unlocked" by some simple hints such as "Try the values of n = 1,2,3,4, and look for a pattern" or "Try drawing a picture." I let the class work on the problems for quite a while, and when they have run out of ideas I offer the hints. With the hints (which are general suggestions, and not problem specific), the students solve the problems in 3 minutes—and they are convinced that I have something to teach them. This is enough to induce them to try the strategies I introduce throughout the term. (Of course, I also tell them that these are specially chosen problems; very few problems can be solved so easily by the application of just one strategy.)

Self-awareness is a more difficult idea to get across. After all, these students are mathematically talented; they're in college precisely because their thinking habits have enabled them to be successful. Yet here I am telling them that they're not very efficient thinkers at all and that they could do a lot better. They won't accept this statement at face value, and they shouldn't: The burden of proof is on me. One way I make the case convincingly is to show the students videotapes of other students working problems, for example, the videotape that produced the graph in Fig. 8.1. The students' reaction is interesting. On one hand, they become quite upset looking at the tapes: "What they're doing is stupid! They're wasting all that time, and it won't do them any good!" On the other hand, they are embarrassed because they empathize with the students: "That could be me." That's precisely the point. It's a lot easier to analyze behavior when it's someone else's, and then to see that the analysis applies to yourself. As a result of watching these tapes and discussing them, the students are made aware of metacognitive issues. They are then more receptive to some of the more interventionist techniques I use later in the term.

Second Technique: Teacher as Role Model for Metacognitive Behavior

Like all teachers, I spend a fair amount of time presenting problem solutions at the blackboard—although I try to keep my "straight" presentations to a minimum, to make time for the alternative class formats to be described below. When I do present solutions, however, I try to do more than just demonstrate the "right" way to get an answer.

When we write the solution to a problem on the board in class, we usually present the results of our thinking in a neat and clean presentation of the answer that we've worked out. In fact, presenting things neatly is part of our professionalism. We may work for quite a while to solve a difficult problem, but the idea is to "get it right" and then present a polished product. Unfortunately, this professionalism has an unintended byproduct. In presenting a polished solution, we often obscure the processes that yielded it, thus giving the impression that things should be easy for people who student the subject matter. In consequence, the give-and-take of real problem solving—the false starts, the recoveries from them, the interesting insights, and the ways we capitalize on them, and so on—are all hidden from students. Yet these are the processes that must be brought out into the open.

One way to bring them out in the open is to model them, presenting "problem resolutions" rather than problem solutions. At times I work a problem as though I were working it from scratch, going blow-by-blow through the solution process. That may mean looking at a few examples to make sure that I understand the problem. Then I may make a few tentative explorations, looking for promising things to do. If I generate a few reasonable approaches, I decide among them and pursue one for a while. After a few minutes of working on it, I reconsider ("Am I making reasonable progress? Does this seem like the right thing to do?") and then act accordingly. I may find a solution along the lines I first pursued, or I may have to back off and try something else before succeeding. This continues until I solve the problem, at which point I do a "post mortem" and review the whole solution.[1]

I am the first to admit that this kind of modeling approach is artificial and must be used sparingly, for it wears thin pretty rapidly. Like the videotape sessions described earlier, these simulated problem resolutions are not intended for extended use. Rather, their primary function is to focus students' attention on metacognitive behaviors. Both methods bring

---

[1]For a detailed description of one such problem resolution, see pp. 8–13 of my (1983) *Problem Solving in the Mathematics Curriculum*. For a far more extensive treatment, see John Mason, Leone Burton, and Kaye Stacey's (1982) *Thinking Mathematically*. That book presents an extended discussion of the give-and-take of problem solving, including "mulling" and what to do when you get "stuck."

those behaviors to center stage, highlighting their importance and preparing the students for more direct work on their metacognitive skills. Such work is the focus of techniques three and four.

Third Technique: Whole-Class Discussions of Problems with Teacher Serving as "Control"

The vast majority of time in my problem-solving classes, and as much time as I can arrange in all of my other classes, is spent with students actively working problems and discussing their solutions. I use two formats for these problem-solving sessions, small-group (discussed in the next section) and whole-class.

When the class works as a whole on problems, I take the role of scribe and orchestrator of the students' suggestions. I do not try to guide the students to the correct solutions, based on my knowledge of the mathematics. This is the standard technique used in "Socratic dialogues"—a technique of some value, but not appropriate for the goals I have in mind. Rather, my task is to help the students make the most of what they themselves generate and to help them reflect on how they do it.

Problem sessions begin when I hand out a list of new problems or ask if any of the homework problems merit class discussion. I write the problem to be discussed on the board and ask if there are any suggestions for solving it. Often one student has an "inspiration" and within a few seconds of reading the problem suggests "Let's try X." At times X is reasonable, and at times not. My task is not to say yes or no, or even to evaluate the suggestion. Rather, it is to raise the issue for discussion. If (as is frequently the case at the beginning of the term) the suggestion has come with breathtaking quickness, I often respond as follows: "Before we try this suggestion, is everyone sure he understands the problem?" Typically a number of students respond no, and I ask how we can make sense of the problem. We then do whatever seems reasonable: exploring the problem conditions, drawing a diagram, working some examples, and so on. When we get to the point where we understand the problem, I return to the original suggestion. "All right, what about X? Is that what we want to do?"

If the original suggestion was inappropriate, our discussions often reveal it to be so: When we've made sense of the problem, the suggestion simply doesn't make sense. As a result of our discussions, the student who had suggested X often realizes that it is not useful and retracts the suggestion on his own. When this happens, I step out of my role as moderator to make the point to the whole class: If you make sure you understand the problem before you jump into a solution, you are less likely to go off on a wild goose chase.

Regardless of the appropriateness of the first suggestion (and whether or not it has been retracted), I ask for other suggestions. Typically there are three or four approaches we might try. As moderator, I ask for a discussion of which one we should attempt, and why. The class makes its decision, and we begin carrying it out. The class may have chosen a direction that leads to a solution, or they may have decided to go down what I know to be a blind alley; either one is fine. I make no attempt to lead the class in the right direction, for what matters is that they make their decision reasonably. Once they have, we work on it.

Whether or not things are going well from my perspective, we stop for a midstream review about 5 minutes into the solution. "We've been doing this for 5 minutes or so. Do things seem to be going pretty well? If so, we should continue. But if not, we might want to reconsider." (Note that it is important to ask these questions even if things are going well; if you use them only when things have gone wrong, students learn to interpret them as a prompt to change directions. The idea is that one should *always* be concerned about the progress of a solution and always ready to reconsider if it seems like things have gotten bogged down.) If the class decides to pursue the direction we have been working in, we continue accordingly. If it decides to abandon it, I ask if anything can be salvaged from what we've done so far: "Are there ideas we might want to return to, or things we might want to try if our new approach doesn't pan out?" We continue along these lines until we reach a solution, at which point I step out of the moderator's role to do a debriefing. There I summarize what the class has done and comment on the efficiency of the solution, pointing out where things have gone awry or where the class might have taken advantage of something they failed to exploit. I also discuss other approaches for working the problem that were advanced during the class's attempt to solve it. We often return to these, producing alternative solutions to the problem.

Whole-class problem sessions like the one just described are an obvious vehicle for dealing with issues of self-regulation. In a sense, self-regulation is the focus of these whole-class sessions. My presence as moderator forces the class to focus on control decisions. However, I am not the one making those decisions: The class itself chooses what to pursue.[2] And although this is a class activity, it works very well for individual students. First, the discussions are out in the open and are later analyzed for their efficiency. This provides the opportunity to reflect on self-regulation and how it works. Second, there is a shared burden; all

---

[2]See pp. 42–50 of my (1983) *Problem Solving in the Mathematics Curriculum* for a blow-by-blow description of a whole-class problem session, in which the class took about forty minutes to solve a difficult construction problem.

students work as a team rather than doing all the work themselves. Because I'm carrying out their decisions at the board, the students can concentrate on decision-making. And with the group mind working toward a solution, no individual student is responsible for generating all of the ideas or for keeping track of all the options. Yet the students do participate actively. In essence they are apprentice problem solvers, working in a real problem-solving context but bearing responsibility for only part of the task they are working on. Like apprentices to a craft, they take over more and more of the task as they gain experience and expertise. (For an excellent discussion of the apprenticeship model of teaching, see Collins and Brown, in press.)

Equally important, the whole-class problem session are an excellent context for discussing students' beliefs about mathematics. Some counterproductive student beliefs were mentioned in the first part of this chapter. In the whole-class problem sessions I can pose problems that evoke the beliefs, and then discuss them with the class. One typical student belief, for example, is that if you really understand the subject matter, then any assigned problem can be solved in relatively short order. In my courses I assign problems that may take the class a few days or even a few weeks to solve—and I let them know (a) that it will take us that long to arrive at a solution and (b) that that situation is perfectly natural, and much more the norm than arriving at solutions in just a few minutes. (I also give 2-week take-home examinations, with the warning that any student who waits until the second week to begin working on the problems is likely to find himself in dire difficulty.)

Problems from geometry provide a nice context for dealing with the beliefs about formal systems just mentioned—that proof has nothing to do with discovery or invention and that the form of a mathematical argument is more important than its content. Chapter 1 of Pólya's (1981) *Mathematical Discovery* is an exceptionally rich source of geometric construction problems, and many of the problems we work are taken from there. In a typical problem, students are asked to construct a problem if they are given some of its "parts"—for example, to construct the triangle that has sides **a**, **b**, and **c**, if you are given line segments whose lengths respectively are the lengths of **a**, **b**, and **c**. This problem is trivial, of course, but its solution leads to interesting discussions. I often assign this as the class's first construction problem, and students usually produce a solution within a minute or two. But when I ask them if they're sure they've got the original triangle (what if I laid down **c** first, then marked off **a** and **b**?), it takes a bit of thought. Eventually they argue that all the triangles gotten by constructions, and any other triangle with sides **a**, **b**, and **c**, must be identical—they're congruent. Next I ask the class to tell me how to bisect an angle, and again the class produces the right construction (from memory). But when I ask them why the construction does indeed bisect the angle, the result is often consternation. One year I let the class address this issue for more than half an hour, without any intervention from me whatsoever. Eventually they convinced themselves that the standard construction creates two congruent triangles and that corresponding angles in those triangles were the two "halves" of the angle that they wished to bisect. At that point a student raised his hand and asked, "Are you trying to tell us that congruence is good for something?"

Indeed I was, and the student's question provided the jumping off point for my discussion. But more important than my discussion was the experience that the students had doing the mathematics and reflecting on it. In working the construction problems, the students learned that they could often *derive* needed information, even when working on what were apparently "discovery" problems. For example, if you want to construct the inscribed circle in a given triangle, you can analyze the properties of the circle. When you discover that the center of the circle lies on the bisectors of the three angles of the triangle, you have identified the information that allows you to construct the circle. This example, too, is fairly straightforward. However, others are quite challenging. Problem 1.19 from Pólya's (1981) *Mathematical Discovery* asks you to construct a triangle, given: the length of one side of the triangle, **a**; the radius of the inscribed circle, **r**; and the measure of the angle **A** opposite the side **a**. This problem occupied the class for the better part of a day. In the process of solving it, we discovered a new theorem, and that theorem allowed us to finish the construction.[3] By using deduction in this way, students came to appreciate its value. By virtue of my boardwork and the classroom discussions, they also came to understand that the form of their argument (e.g., whether or not a proof was written in two-column format) was not important: What mattered was that their argument had to be complete, logical, and coherently argued.

Another perspective on what I do in the group context is suggested by Collins and Brown's (in press) notion of apprenticeship. Students enter my courses with a set of beliefs about mathematics, beliefs abstracted from their previous experiences with mathematics. That previous experience was, for the most part, training for mastery: The students were given training in ready-made, prepackaged mathematical procedures. In the traditional sense, apprenticeship does much more than provide training; it provides an initiation into a culture. (Consider, for example, the contrast between (a) taking a journalism class and (b) spending a summer doing

---

[3]For a more extended discussion of the relationship between deduction and discovery, see my (1986) article, "On having and using geometric knowledge."

entry-level work for a newspaper.) When you learn about a discipline through apprenticeship into it, you are less likely to pick up the kinds of incorrect and counterproductive beliefs about it described earlier. Thus, a significant part of what I attempt to do (in my problem-solving courses in particular, but increasingly in all of my mathematics instruction) is to create a microcosm of mathematical culture—an environment in which my students create and discuss mathematics much as mathematicians do. Having experienced mathematics in this way, students are more likely to develop a more accurate view of what mathematics is and how it is done.

Fourth Technique: Problem Solving in Small Groups

About a third of the time, the students in my problem-solving courses break into groups of three or four to work together on problems that have just been handed out. While they are working on the problems, I move from group to group, answering questions and offering advice.

I began this practice some 10 years ago, on the basis of common sense and a metaphor. To begin with, it just didn't seem right to conduct a problem-solving class solely in lecture or recitation mode. Common sense dictated that, at least some percentage of the time, students in a problem-solving course should be actively engaged in solving problems. But at the heart of the matter—and the heart of the metaphor—was how I defined my role as a teacher. In standard courses, teachers tend to think of themselves as purveyors of information: "Here's what's known, and here's how it's used." Of course delivering such information was a component of my problem-solving courses. I demonstrated various heuristic strategies, for example. But a large part of my task was to help the students become good problem solvers, by helping them to make efficient use of what they already knew. In that sense, my task was more like that of a coach, an "intellectual coach" to be sure, but one who spends a lot of time working with students to help them improve their problem solving ability.

The coaching metaphor cast my instruction in a different light. Sports coaches spend a fair amount of time showing their students "how to do it right," of course. A tennis coach will demonstrate a good serve, a swimming coach the right way to hold one's arm for the butterfly, and so forth. But coaches who stop at that won't keep their jobs for very long. (Imagine a swimming coach who demonstrates a particular stroke and then says, "Your homework is to practice this stroke. There will be a test on Friday before the meet.") Coaches watch students as they practice and as they compete. They make corrections "on line," because this is the context in which the coaching will have an effect. Indeed, coaches videotape their students as they practice and go over the tapes with the students in slow motion replays, because that kind of microscopic analysis can help reveal what's working and what's going wrong.

I had a similar responsibility to my students. My goal was to provide them with a variety of problem-solving techniques (various heuristics) and then to coach them in the efficient use of those techniques. If I showed the students the techniques and then said "Go home and practice," I would be missing the opportunity to intervene when it mattered most—when the students were in the midst of problem solving. Indeed, lectures on "monitoring your solution" and "not going off on wild goose chases" are almost guaranteed to have no effect unless there is in-class problem solving. Students who hear me talk about self-regulation but then go back to their dormitory room to do their homework will almost certainly fall right back into their old habits. Such old habits die hard (especially if the students have been successful with them!), and a few words of warning from a teacher are unlikely to change them.

In fact, the students' behavior is unlikely to be affected unless, at least at the beginning, the coaching is fairly pointed. In handouts and my introductory discussion at the beginning of the course, I inform the students that I reserve the right to ask them the following questions[4] at any time during their problem solving:

- What (exactly) are you doing? (Can you describe it precisely?)
- Why are you doing it? (How does it fit into the solution?)
- How does it help you? (What will you do with the outcome when you obtain it?)

About 2 weeks into the course, I start asking the small groups these questions as I move through the room in my role as problem-solving consultant. I do so gently, reminding the students that I had warned them about this aspect of the course. At first the reaction from each small group is an embarrassed silence. (More precisely, the students can usually answer the first of the three questions, but they have no answers for the second and third.) With apologies, I persevere.

Soon the students realize that I'm serious about the questions and that I will continue to ask them even though doing so makes them feel uncomfortable. To defend themselves against these intrusions, they begin

---

[4]These questions might leave the impression that I expect the problem solver always to be "on task," knowing exactly what he's doing and why. That's not the case. I'm not, and I don't expect my students to be. One perfectly good answer to the first question, for example, is "I'm mucking around looking for inspiration, and I intend to do so for another 5 minutes." As long as the student is aware of what he's doing, the odds are that he won't continue mucking around indefinitely.

to prepare answers to the questions in advance. Over the course of the semester, the students get in the habit of discussing the questions, both at the beginning of problem sessions and at major decision points during problem solutions. When things work well, discussions of the underlying issues—whether what the students are doing is reasonable and whether they are making progress—become a matter of practice, and there is less and less need to invoke the questions themselves.

I need to back off for a moment. The preceding discussion may be laying undue emphasis on those questions and, correspondingly, on the way that the questions, and my interference, shape the group dynamics. In fact, each group of students functions on its own about 95% of the time. In a class of 20–30 students, there are usually between six and eight working groups. At best (or worst, depending on your point of view), I only have between one sixth to one eighth of the time to spend with any group. In practice, I spend less time. By the middle of the term, the group dynamics are well established. I hand out a new set of problems, and the class gets to work. Often I return to my desk, or leave the room for a few minutes. If a group requests my help as a consultant, I work with them for a while.[5] Then I move to another group, watching their work for a while and asking if things are going well. If the answer is yes, I move on. Often I return to my desk, and the groups continue working alone until I get the feeling that they've reached closure. Then we convene as a "committee of the whole" to discuss the progress they have made.

Does all of the attention to metacognitive issues in my courses pay off? One must be careful in answering such a question, because it's not always clear what pays off in what ways. That is, a course may produce certain kinds of results, but it's dangerous to attribute particular results to any particular aspect of the course. (Indeed, if you take the "cultural transmission" point of view explored later, it may be impossible to do so.) But there is clear evidence that the course produced marked shifts in the students' problem-solving behavior, particularly at the metacognitive level.

I reported earlier that more than half (actually closer to 60%) of the problem-solving sessions I have analyzed are of the "read the problem, make a decision to do something, and then pursue it come hell or high water" type illustrated in Fig. 8.1. After taking my problem-solving courses, fewer than 20% of my students' solution attempts are of that

---

[5]As the term progresses there are fewer requests for my assistance. This happens both because of acculturation and because of the evolution of the group dynamics. First, the students come to realize that they are not engaged in a contest where the goal is to get the right answer and move on. Their goal (supported by the environment) is to make sense of the mathematics. Second, they realize that it's much better, and more fun, for them to make sense of the problems by themselves.

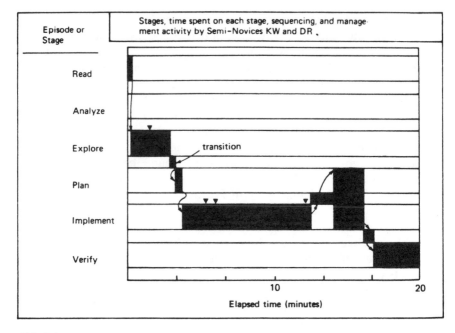

FIG. 8.4.

type. Figure 8.4 presents the graph of a relatively typical postinstruction attempt to solve a nonstandard problem. The entire solution, and a discussion of it, may be found in chapter 9 of Schoenfeld (1985). What follows is a brief outline of their attempt.

As Fig. 8.4 indicates, the students solved the problem they were asked to work. That's a nice bonus, but it's not central to the current discussion. What is important is the route the students took along the way to getting their answer. They began typically, reading the problem and then jumping into a solution attempt.[6] But 2 minutes into the solution, one of the students commented that they might be barking up the wrong tree and that the attempt they were making was based on a possibly unfounded assumption. They stopped what they were doing and decided to try something else. Unfortunately, what they decided to try was also wrong. For 8½ minutes, they were involved in complicated calculations, which ultimately led nowhere. But about 12 minutes into the solution, they pulled up short once again. "No, we aren't getting anything here. We're just getting . . . [what we're doing is unjustified] . . . Let's start all over and forget about this." Within the next 2 minutes they found an approach

---

[6]As I mentioned earlier, old habits die hard.

that worked, solved the problem, and confirmed their solution's correctness.

Viewed from one perspective, the way this problem was solved was hardly "ideal." The students began with a leap into exploration, which could have been avoided. Soon after that they embarked on an $8\frac{1}{2}$-minute-long wild goose chase that, like the first exploration, could have been forestalled if they had carefully examined what they planned to do. There is still a fairly sharp contrast between this graph and the graph of an expert's problem solution shown in Fig. 8.2. But to make such comparisons is unfair and misses the point. The real comparison is with the graph in Fig. 8.1.

As noted, the students did solve the given problem. Since solving nonstandard problems can be a function of luck and prior knowledge, the fact that they succeeded must to some degree be considered a happy accident; one cannot take credit for this aspect of their success. What is not accidental, however, is that the students had the *opportunity* to find their solution. If the problem session had been recorded before the course, there is a good chance that the students would have pursued their first guess for all of their allotted time. (Nearly 60% of their peers did, and they themselves did on problems recorded before the course.) And if they had escaped that trap, another and much more serious one was waiting for them: The deatiled computations that occupied them for $8\frac{1}{2}$ minutes. In my experience, students who become that deeply immersed in computations virtually never manage to extricate themselves from them. It is almost certain that, before the course, poor metacognitive skills would have guaranteed their failure. After the course, the possession of such skills could not guarantee success, but at least it could give them a fair shot at it. By avoiding one wild goose chase entirely and bringing the second to a halt, the students salvaged their solution attempt twice. The result was that they then had the opportunity to find an approach that solved the problem (and they did so). Viewed from that perspective, their solution was indeed expertlike. In this and in other ways, the instruction has been quite successful.

IN SUMMARY, A SEMITHEORETICAL COMMENTARY

The instructional techniques described in this chapter have been justified by common sense, a metaphor, and one powerful post hoc argument— they work. In this concluding discussion I would like to examine some of the reasons why they work. What follows is a speculative exploration of the role of social context in the development of metacognition. It is intended more to establish an agenda of interesting items to think about than to say anything definitive.

One way to characterize efficient self-regulation is to say that the people who are good at it are the people who are good at arguing with themselves. In the transcript of problem solving session graphed in Fig. 8.2, for example, you can almost see the problem solver taking on different roles as he works the problem. At different times, he functions as an idea generator, a systematic planner, a critic, a "progress monitor," an advocate first for one particular point of view ("Maybe I should do it this way . . .") and then for another, and so on. To use a popular phrase from cognitive science, you see him function in his problem solving as a "society of mind"—putting forth multiple perspectives, weighing them against each other, and selecting among them. When this works well, it is highly productive. The idea generator suggests many interesting things for him to try as he works, while the critic and monitor can keep him from going off on wild goose chases.

Where do these skills come from? One point of view, pioneered by the Soviet psychologist L. S. Vygotsky in his *Thought and Language* (1962) and *Mind in Society* (1978), is that all higher order cognitive skills originate in, and develop by the internalization of, individuals' interactions with others.

> Every function in the child's development appears twice: first, on the social level, and later on the individual level; first, *between* people *(interpsychological)*, and then *inside* the child *(intrapsychological)*. This applies equally to voluntary attention, to logical memory, and to the formation of concepts. All the higher functions originate as actual relationships between individuals. (Vygotsky, 1978, p. 57)

Vygotsky hypothesizes that the potential for development at any time is limited to what he calls the "zone of proximal development (ZPD)," defined as follows. Working alone, the child may function up to a certain level. Working in collaboration with more capable peers, or perhaps with adult guidance, the child may function at a somewhat higher level. This middle ground, which the student is capable of reaching with some assistance but not on his own, is the ZPD. Vygotsky's thesis is that one acquires higher order skills by exercising those skills in the ZPD with the help of others and then internalizing those skills, that is, by mastering as an individual those skills for which one, at one time, needed support.

This perspective provides strong justification for the use of small groups in problem-solving contexts. Suppose the description of the good problem solver as one who argues (intelligently) with himself is right and that a large component of effective problem solving consists of advancing multiple perspectives, balancing them against each other, and proceeding on the basis of what seems, on balance, to be the best option at the time. Almost by definition, small-group discussions (when they work well) result in the individual's working in his ZPD. While the individual might generate one possibility and go off in pursuit of it, a group might generate three or four—and precisely because they have more than one option, the group will have to decide among them. In consequence, the individual might have to formulate and defend one point of view, listen to and evaluate others, and finally take part in a group decision regarding which one(s) to pursue and for how long. These are precisely the self-regulation skills the individual needs to develop, and it is difficult to imagine a context in which they could develop more naturally.[7]

For some years this version of the Vygotskean hypothesis has been the primary theoretical support for my use of small groups. As noted earlier, the coaching metaphor provided additional justification. For the sake of completeness, here are two other arguments I (1983) have used:

> Problem solving is not always a solitary endeavor. Students have little opportunity to engage in collaborative efforts, and this does not do them any harm.
> Students are remarkably insecure, especially in a course of this nature. Working on problems in groups is reassuring: one sees that his fellow students are also having difficulty, and that they too have to struggle to make sense of the problems that have been thrown at them. (pp. 30–31)

I have come to think that such statements fail to do justice to the phenomena we have been discussing. The issue is much larger than these arguments would suggest, and it is primarily a cultural one.

Let us return to the topic of collaboration. The stereotype is of the mathematician alone in his office, scrawling on a pad long into the night. For this reason it is especially interesting to note how mathematicians themselves discuss the topic. Donald Albers' and Gerald Alexanderson's recent (1985) *Mathematical People: Profiles and Interviews* offers a collection of conversations with and about contemporary figures in mathematics. It is surprising how frequently the mathematicians mention collaboration, both in terms of influences on their work and in terms of the benefit they derive from working with others. In one case, "collaborative distance" has actually been formalized. Paul Erdös is a remarkably

---

[7]I have thought about slightly less natural small group interactions, designed to promote the growth of particular characters in the society of mind. Suppose you structured small groups so that individuals had (one or more) assigned roles—say as idea generator, planner, critic, monitor, etc. Making these role assignments overt might make the tasks more salient, and provide more direct practice at them. This seems like an interesting idea, but I haven't tried it. I suspect that making the right role assignments and getting the group dynamics to work might be decidedly non-trivial.

prolific mathematician who has produced some 900 papers, of which about 200 were coauthored. Each of his coauthors has an "Erdös number" of 1. Each of the people who has coauthored a paper with one of them has an Erdös number of 2, and so on. Einstein had an Erdös number of 2, and there is ongoing historical research to determine the Erdös number assigned to C. F. Gauss (1777–1855), one of the greatest mathematicians of all time.

Peter Hilton lays out the benefits of collaboration as follows (Albers & Alexanderson, 1985):

> First I must say that I do enjoy it. I very much enjoy collaborating with friends. Second, I think it is an efficient thing to do because . . . if you are just working on your own [you may] run out of steam. . . . But with two of you, what tends to happen is that when one person begins to feel a flagging interest, the other one provides the stimulus. . . . The third thing is, if you choose people to collaborate with who somewhat complement rather than duplicate the contribution that you are able to make, probably a better product results. (p. 141)

The second and third reasons are rational and important, but Hilton lists the joy of collaboration first. Persi Diaconis says the following (Albers & Alexanderson, 1985):

> There is a great advantage in working with a great co-author. There is excitement and fun, and it's something I notice happening more and more in mathematics. Mathematical people enjoy talking to each other. . . . Collaboration forces you to work beyond your normal level. Ron Graham has a nice way to put it. He says that when you've done a joint paper, both co-authors do 75% of the work, and that's about right. . . . Collaboration for me means enjoying talking and explaining,, false starts, and the interaction of personalities. It's a great, great joy to me. (pp. 74–75)

The last quotation captures much of the essence of what it is to be a mathematician, or, better, to be a member of the mathematical community. Diaconis speaks of excitement, of mathematical people (note the very phrase!) taking pleasure in talking to each other; of talking and explaining, of the joy in false starts, and the interaction of personalities. Not only is it a "great, great joy"; I would argue that those interactions, and the sense of community—a culture of mathematics, if you will—are part of what sustains mathematics. And participation in that culture is how one comes to understand what mathematics is.

I remember discussing with some colleagues, early in our careers, what it was like to be a mathematician. Despite obvious individual differences, we had all developed what might be called the mathematician's point of view—a certain way of thinking about mathematics, of its value, of how it is done. What we had picked up was much more than a set of skills; it was a way of viewing the world and our work. We came to realize that we had undergone a process of acculturation, in which we had become members of, and had accepted the values of, a particular community.[8] As the result of a protracted apprenticeship into mathematics, we had become mathematicians in a deep sense (by dint of world view) as well as by definition (what we were trained in and did for a living).

Now, what does this have to do with ordinary, day-to-day mathematics instruction? I think a great deal. If the notion of culture as explored here makes sense, it explains why my problem-solving courses have been successful in the way that they have been—and why so many attempts at curriculum reform have failed.

My problem-solving courses have evolved in an interesting way. Early versions of the courses focused on the thinking tools I thought students needed for competent problem-solving performance. This meant giving training in particular heuristic strategies, teaching a prescriptive "managerial strategy" for self-regulation, and pointing directly to problems (or domains) where the "wrong" beliefs caused difficulty. Over time the managerial strategy disappeared (it was artificial) and was replaced by the more natural in-class techniques described above. Work on beliefs moved in the same direction. I began with "damage containment," identifying counterproductive beliefs and dealing with them on a case-by-case basis. Later I moved toward a natural environment, in which we worked as a group doing mathematics the "right" way.

With hindsight, I realize that what I succeeded in doing in the most recent versions of my problem-solving course was to create a microcosm of mathematical culture. Mathematics was the medium of exchange. We talked about mathematics, explained it to each other, shared the false starts, enjoyed the interaction of personalities. In short, we became mathematical people. It was fun, but it was also natural and felt right; it wasn't a separate "school experience" for a few hours a week. By virtue of this cultural immersion, the students experienced mathematics in a way that made sense, in a way similar to the way mathematicians live it. For that reason, the course has a much greater chance of having a lasting effect.

---

[8]Let me give just one example. To this day I have the vivid memory of Peggy Strait, who taught my undergraduate probability course at Queens College, going to the board to write the statement of a theorem. When she got to the statement of the main result she stopped and said "I never remember this result, but that's no problem; it's so easy to derive." She did just that, showing us why it made sense; then she finished writing the statement of the theorem. At that moment I saw how mathematics should be: if you really understand it, you don't have to memorize a lot, because you can figure it out. This became part of my sense of what mathematics is all about.

I argued earlier that the interactions among mathematicians, and the sense of community they support, are part of what sustains mathematics—that the practice of mathematics is a human endeavor and very much a cultural one. I also suggested that entry into that culture (or some culture that supports the same values) may be necessary to understand and appreciate mathematics. If this is right, it explains at least partially the failure of a whole slew of curriculum reform movements—moves for "relevance," for the "new math," for "basic skills," and, I predict, for "problem solving." Each of these curriculum reforms reflects an attempt to embed a selected aspect of mathematical thinking into what is an essentially alien culture, that of the traditional classroom. As long as the two cultures differ as radically as they do at present, it may be impossible for this kind of embedding to succeed. Fragments of the mathematical culture, in isolation, are likely to wither in the classroom for lack of support, or they will be so changed by their absorbtion into the classroom culture that they will not transfer back outside of it. This holds even for attempts to link schooling and the real world, such as the "relevance" movement and attempts to use "applied" problems in curricula. According to this view, those attempts didn't have a chance.

Although these comments may sound pessimistic, my sense is just the opposite—that substantial optimism is warranted. If there is real substance to this notion of culture as the medium for knowledge transmission, then there is the potential for a research program that may lead us out of the "school learning doesn't transfer to real life" dilemma. The idea, of course, would be to engender a culture of schooling that reflects the use of mathematical knowledge outside the school context. To achieve this we need to understand more about mathematical thinking and culture in at least two ways. In this chapter I have focused on elaborating aspects of mathematical thinking as seen from the mathematician's point of view—what it is to live in a culture of mathematics. But the flip side of the coin, understanding the development of mathematical thinking in our culture at large, is equally crucial. There is a growing body of anthropological work on cognition in practice (see, e.g., Carraher, Carraher, & Schliemann, 1985; Lave, in press; Rogoff & Lave, 1984) indicating that people are very good at inventing the mathematics they need to carry out tasks that are really important to them. My sense is that the two cultures are compatible and that, when we understand enough about them, they can be reasonably conjoined. If this is the case, what we need is a program of "cultural design" for schooling. Understanding enough about the social contexts that promote the need to develop and understand mathematical ideas, and about the environments that support the growth and development of those ideas, may allow us to create classroom cultures in which students *do* mathematics naturally. When that happens, the "transfer problem" will no longer be a problem.

REFERENCES

Albers, D. J., & Alexanderson, G. L. (1985). *Mathematical people: Profiles and interviews.* Chicago: Contemporary Books.

Brown, A. L. (1978). Knowing when, where, and how to remember: A problem of metacognition. In R. Glaser (Ed.), *Advances in instructional psychology* (pp. 77–165) (Vol. 1). Hillsdale, NJ: Lawrence Erlbaum Associates.

Brown, A. L., Bransford, J., Ferrara, R., & Campione, J. (1983). Learning, remembering, and understanding. In P. H. Mussen (Ed.), *Handbook of child psychology* (Vol. 3) (pp. 77–166). New York: Wiley.

Carpenter, T. P., Lindquist, M. M., Matthews, W., & Silver, E. A. (1983). Results of the third NAEP mathematics assessment: Secondary school. *Mathematics Teacher, 76*(9), 652–659.

Carraher, T. N., Carraher, D. W., & Schliemann, A. D. (1985). Mathematics in the streets and in the schools. *British Journal of Developmental Psychology, 3,* 21–29.

Collins, A. L., & Brown, J. S. (in press). The new apprenticeship: Teaching students the craft of reading, writing, and mathematics. In L. B. Resnick (Ed.), *Cognition and instruction: Issues and agendas.* Hillsdale, NJ: Lawrence Erlbaum Associates.

Lave, J. (in press). *Arithmetic practice and cognitive theory.* Cambridge: Cambridge University Press.

Mason, J., Burton, L., & Stacey, K. (1982). *Thinking mathematically.* New York: Addison-Wesley.

Pólya, G. (1981). *Mathematical discovery* (Combined paperback edition). New York: Wiley.

Rogoff, B., & Lave, J. (1984). *Everyday cognition: Its development in social context.* Cambridge, MA: Harvard University Press.

Schoenfeld, A. H. (1983). *Problem solving in the mathematics curriculum: A report, recommendations, and an annotated bibliography.* Washington, DC: Mathematical Association of America.

Schoenfeld, A. H. (1985). *Mathematical problem solving.* Orlando, FL: Academic Press.

Schoenfeld, A. H. (1986). On having and using geometric knowledge. In J. Hiebert (Ed.), *Conceptual and procedural knowledge: The case of mathematics.* Hillsdale, NJ: Lawrence Erlbaum Associates.

Schoenfeld, A. H. (in press). When good teaching to bad results: The disasters of well taught mathematics courses. *Educational Psychologist.*

Vygotsky, L. S. (1962). (E. Hanfmann & G. Vakar, Eds. & Trans.). *Thought and language.* Cambridge, MA: MIT Press and Wiley.

Vygotsky, L. S. (1978). (M. Cole, V. John-Steiner, S. Scribner, & E. Souberman, Trans.) *Mind in society.* Cambridge, MA: Harvard University Press.

9 Cognitive Science and Algebra Learning

Ronald H. Wenger
University of Delaware

This chapter is concerned with the applications of research in cognitive science to practical instructional issues in algebra. After this brief introduction, the chapter is divided into two main parts. The first describes several projects in cognitive science that I have found particularly useful in my efforts to understand students' difficulties with algebra and to remediate them. The main questions addressed in the first part are the following: (a) Do attempts to model mathematics learning via computer programs have useful implications for mathematics education? (b) What can we learn from the methods used by researchers in Artificial Intelligence to construct symbolic algebra environments on computers?[1] The second part explores the practical applications of the research described in the first part. I hope to show that cognitive research has direct implications for everyday classroom practice. The main questions addressed in this part are the following: (a) Does information from cognitive science regarding the ways that students learn suggest that we might develop different objectives for the curriculum? (b) Does information from cognitive science suggest ways of designing more effective instructional strategies and materials?

A brief description of my background will help to explain the perspective taken in this chapter. I am a mathematician, not a cognitive scientist. As a teacher of mathematics I became concerned with students' difficulties in learning some of the more formal aspects of algebra and elemen-

---

[1]Note that this question is related to, but quite different from, the question "How can computer-based algebra environments be used to shape the curriculum or improve instruction?"

tary functions. I saw conscientious college students who, having serious difficulties with the subject matter, worked quite hard at trying to learn or relearn college algebra. Were there some not obvious reasons for these difficulties? I wondered if there might be something wrong with the way we organized the subject matter or the way we presented it. I wondered if we might be able to make efficient instructional use of some of the new computational tools I had encountered as a mathematician, for example the powerful symbolic algebra computing environments (such as mu-Math[2]) now available on microcomputers. As I looked for help in dealing with these questions, I discovered that research from cognitive psychology and artificial intelligence (two of the fields that have contributed to cognitive science) was quite informative. That research[3] provided my colleagues and me with ways of analyzing our students' difficulties with algebra and suggested instructional approaches that might help to avoid some of those difficulties. It also suggested ways that symbolic algebra environments might be profitably used for instruction and provided some theoretical principles useful for the development of computer-based algebra tutoring systems.

In this chapter I discuss the literature that we found useful, show how that literature affected our thinking about the ways that students learn—or have difficulties learning—algebra, and discuss the implications of that literature for curricular and instructional practice. In the next few pages I describe some background issues: some typical student difficulties that motivated our work, some general ideas from cognitive science that shaped our inquiries, and some specific assumptions about students' learning of algebra that frame the discussion in this chapter.

SOME SYMPTOMS OF DIFFICULTY AND SOME BASIC ASSUMPTIONS

One student difficulty that motivated much of our work is familiar to all teachers of algebra. Our students, many of whom had extensive experience with algebra in high school, worked hard, learned the techniques in the various chapters of their textbook, and were able to solve the

---

[2]muMath is a symbolic algebra program that runs on microcomputers. It successfully carries out many of the procedures currently emphasized in high school and college mathematics curricula; see Heid (1983) for a discussion of its use in a calculus course. The version of muMath discussed in this paper runs on the IBM PC and is available from The Soft Warehouse, P.O. Box 11174, Honolulu, Hawaii.

[3]The emphasis here is on cognitive science research. But there is an extensive literature on algebra that predates cognitive science or for which this perspective is not central; for example, the work of Booth (1984); Kieran (1979a, 1979b), 1981); Wagner, Rachlin, & Jensen (1984).

problems in the individual chapters. Yet they did poorly on tests that covered several chapters at a time. The students could use the various techniques they had studied, but when problems were presented out of context and the students had to select the methods as well as use them, they had great difficulty.

A second problem was that students just didn't seem to "see" the right things in algebraic expressions or in the equations they were asked to solve. Consider the following problem, which will be used as an example throughout the chapter:

Equation #1: Solve the equation $v\sqrt{u} = 1 + 2v\sqrt{1+u}$ for v.

If you can see your way past the morass of symbols and observe that equation #1 is linear in v, the problem is essentially solved: an equation of the form

$$av = b + cv$$

has a solution of the form $v = b/(a-c)$, if $a \neq c$, no matter how complicated the expressions a, b, and c may be. Yet students consistently have great difficulty with such problems. They will often perform legal transformations on the equations, but with the result that the equations become harder to deal with; they may "go round in circles" and after three or four manipulations recreate an equation that they had already derived; they often appear to choose their next move almost randomly, rather than with a specific purpose in mind. Note that in these examples the students sometimes perform the manipulations correctly; they may not make what are typically called "procedural errors" in manipulating the equations. Rather, they make what I will call "strategy errors"—poor choices of what to do next. A major focus of this chapter is on such strategy errors. Our goal is to understand the sources of such errors and find ways of remediating them (or, better yet, helping students to avoid them).

Some of the major themes underlying our work have already been discussed in this book. An important aspect of our approach is a constructivist point of view and a commitment to the concept of students as interpreters of their experience rather than as absorbers of knowledge (see chapter 1). That being the case, we ask if some of the unproductive behavior patterns that concern us—student difficulties with "literals" or parameters in expressions, their inability to see and exploit patterns in equation problem solving, their lack of planning, and so on—might be the inadvertent result of current textbook design. Does the packaging of subject matter into single-technique chunks in a one-rule-per-section format leave students with the impression that solutions should be straightforward, requiring only the use of one rule or procedure? Does it give students the impression that mathematical problem solving is based on an arbitrary collection of rules, so that planning does not seem to be a natural activity and hence they don't plan?

A second major theme deals with knowledge organization. As mentioned earlier, students can often perform the required symbolic manipulations correctly, but they have difficulty knowing which approaches to select. This observation suggests that students' difficulties result not so much from the content of their mathematical knowledge base but from its organization. In a discussion of planning, Sacerdoti (1977) expresses a similar point of view. "If the results of the work described in this monograph can be described in a phrase, it is this: the structure of knowledge about actions within a computer memory is as important as the content of that knowledge" (p. 105). As we have seen throughout this book, that comment reflects a fundamental assumption that cognitive science makes about human planning, problem solving, and thought processes in general. What one knows is only part of the story. The ways in which that knowledge is organized, accessed, and used are equally important determinants of one's intellectual performance. For detailed discussions of the role of knowledge organization in various problem-solving domains, the reader may wish to explore Anderson (1983), Reif (1983), Sacerdoti (1977), and Schoenfeld (1985).

The third major theme was introduced in chapter 1. In brief, we focus on process rather than product—on what students do, rather than the results they produce. This work is done at a very fine level of detail. As Brown and Deloache (1978) state, "The main emphasis [of cognitive science] is on providing detailed explicit models of cognitive development within a limited task domain. The aim is to provide precise descriptions of the initial and final forms of the cognitive process under investigation and to delineate important intermediate stages" (p. 8). The process models we use are detailed and complex. They apply an "information processing" perspective to human cognition, and make certain basic assumptions about the "structure of mind," including assumptions about working memory, long-term memory, control processes, and so on. Ed Silver (chapter 2) provided a basic introduction to some of these ideas. Other introductions to this area can be found in Resnick and Ford (1981, ch. 8), Davis (1984, chs. 4 & 5), and Gentner and Stevens (1983, ch. 1). Anderson (1981) contains a more detailed and research-oriented discussion.

With these general comments as background, I would like to focus on a specific set of assumptions relevant to the balance of this paper. While they may well be generalizable to other age groups, these assumptions stem from our work with high school or college students enrolled in a course in algebra or elementary functions.

Assumption 1: Students learn mathematics primarily from examples and practice tasks—typically from textbooks.[4] That is, they do not usually learn by understanding the explanations of procedures and using those explanations. Rather, they figure out what the procedures are about by working through them.

Assumption 2: An important force driving student behavior is the need to make sense of things by creating simple, straightforward procedures that work. As students work problems, they invent rules for the procedures that seem to fit the expected answers. Often those rules are as simple as possible, and often they are incorrect.

Assumption 3: The rules and procedures that students invent become part of their problem solving approach. A student will continue to use one of these inferred procedures until (at least) encountering a task that reveals it as being incorrect.

If one accepts these assumptions, one sees immediate instructional consequences.

Instructional Consequence 1: If a list of textbook tasks can be completed successfully (with the right answers) using a sequence of procedures inferred by the student, then the correctness or the validity of the strategy is confirmed and the student believes he or she "understands" the topic.

Instructional consequence 1 applies whether or not the students' inferred strategies and procedures are valid. If a student can solve a set of sample problems (get the right answer) using a procedure simpler than the correct one, the student may believe that simpler procedure *is* the correct one. If problems are organized so that some incidental feature of the problems become salient to the student, the student may come to believe that that incidental feature is central to the procedure. Thus, the nature and sequence of the problems that students work are critically important in the understandings that the students develop. To borrow Steve Maurer's phrase, these two examples of student mislearning describe the bad news about assumptions 1 through 3. However, there is good news as well:

Instructional Consequence 2: Our primary leverage for instruction lies in making the forces described in the assumptions work for us, by the careful selection and design of the examples and of the sets of practice tasks that students will work.

---

[4]The work of VanLehn (1983) concerning felicity conditions, although not discussed here, is quite important. While these assumptions are our own observations, they are quite consistent with the conditions he discusses; see, for example, p. 10.

SOME RELEVANT PROJECTS IN COGNITIVE SCIENCE
AND ARTIFICIAL INTELLIGENCE

One of the primary benefits of cognitive science research for mathematics learning is that it provides a language in which important questions of long standing can be described together with theories that provide ways of systematically addressing these questions. Many of the findings of this research have immediate implications for instruction, though this was not always the case with earlier normative statistical methodologies. The projects described here illustrate these implications. They also strongly influenced the perspectives and methodologies on which the applications discussed later are based.

The Debuggy Metaphor

Buggy, debuggy, repair theory and the continuing work of Brown and Burton (1978) and VanLehn (1981, 1983) at Xerox PARC have influenced us significantly even though these studies concerned topics in arithmetic, whereas our interests lie in algebra and elementary functions. Since Steve Maurer's chapter gives a descriptive overview of this work, we need not do so here. Instead, we shall describe how we have been attempting to extend the "debuggy metaphor" to higher mathematical levels. The assumptions discussed in the introduction to this chapter are a part of that effort. We have found that the debuggy metaphor yields useful information about students' algebraic behavior, both valid and errant, on topics in algebra and elementary functions.

As indicated in chapter 1, cognitive science sees learning as a process and studies the relationship between that process and hypothesized cognitive models. It is this perspective and the resultant focus that cognitive science offers the mathematics teacher. The buggy/debuggy work is prototypical. Although analyzing student behavior in the classroom was not its original purpose, it provides an extremely powerful and constructive way of thinking about and observing that behavior. It also has important implications for the design of instructional materials and related sets of practice tasks. These implications are apparent in the reaction of one gifted and experienced teacher to papers describing this work. She wrote, "Explanations of the thought processes of the students and the origin of many of the errors has already given me some insight on how to change my instruction." This is a not uncommon response.

The essence of the debuggy metaphor is that there are identifiable and systematic underlying causes of students' errant behavior and that these causes often have deep and pedagogically powerful implications. The causes of errant behavior may be, in part, the very specific ways in which students understand or misunderstand a task. The errors, therefore, are not just errors, but "bugs" symptomatic of the form of the students' understanding. As in the buggy work it is clear that to focus on the errors and attempt to treat them individually is pedagogically unwise. It is also virtually impossible, at least in algebra, because of their great numbers. Although this specific approach is new, arithmetic errors, algebraic errors, and the concept of variable and equation have been studied by mathematics educators for many years. A sample of their work and an extensive bibliography are contained in studies by Booth (1984). Similar questions are approached from an information-processing perspective by Carry, Lewis, and Bernard (1981), Clement, Lockhead, and Soloway (1980), and Matz (1980, 1982).

The work of Matz (1980, 1982) is an effort to extend some of these themes to algebra. She is interested in the causes of certain algebraic errors such as the ubiquitous linearity errors of the form $f(A + B) \rightarrow f(A) + f(B)$ for functions such as $f(x) = x^2$ or the errors illustrated for radical equations later in the paper (e.g., Fig. 9.3). She suggests "that errors are the results of reasonable, although unsuccessful, attempts to adapt previously acquired knowledge to a new situation" (Sleeman & Brown, 1982, p. 25). The student often extracts the procedure in the form of a rule from a prototypical example or directly from a textbook.

In the second part of this chapter these ideas are extended to "strategic errors" in the context of tasks like solving equation #1. To anticipate the discussion there, an example of an annotated protocol[5] for one student's work on Equation #1 is provided in Fig. 9.1. The left-hand column represents the series of steps taken by the student; the right-hand column contains the researcher's comments. Note the strategic error from step 2 to step 3.

The student was a high-school senior but, as the discussion later in this chapter will show, these kinds of errors are not limited to high-school students. When college freshmen were given the problem in a precalculus couse, more than a third of them developed similar errant strategies. Our interest is in students' "strategic" and "planning" behaviors on such tasks. Consequently, the protocol does not dwell upon their detailed manipulative steps. The protocol also contains brief annotations providing hypothesized purposes or descriptions of the action taken to transform each equation into the next. In effect, these annotations represent the perspective of the "skilled" equation solver.

Although Matz (1982) did not study errors in strategies for equation

---

[5]Protocols are a primary source of "data" used by cognitive scientists to study such behavior. Schoenfeld (1983) describes them as: "objective records or traces of a sequence of overt actions taken by individuals in the process of solving problems" (p. 347).

1) $v\sqrt{u} = 1 + 2v\sqrt{1 + u}$ (equation statement)

[1,2]   Transformation: Add -1 to both sides.
Purpose: Appears to prefer to isolate the more complex radical term before applying the square-both-sides procedure.

2) $v\sqrt{u} - 1 = 2v\sqrt{1 + u}$

[2,3]   SERROR: Strategic ERROR of applying the square-both-sides procedure. Appears not to notice expressions are linear in goal variable, v.

3) $(v\sqrt{u} - 1)^2 = (2v\sqrt{1 + u})^2$

[3,4]   Rewrite rule: $A^2 \rightarrow A*A$

4) $(v\sqrt{u} - 1)(v\sqrt{u} - 1) = (2v\sqrt{1 + u})^2$

[4,5]   Rewrite rule: expand $A*A$ on 1hs (left-hand side), square monomial on rhs.

5) $v^2u - 2v\sqrt{u} + 1 = 4v^2(1 + u)$

[5,6]   Pattern: clear-paren
SERROR: Better to have left factored.

6) $v^2u - 2v\sqrt{u} + 1 = 4v^2 + 4v^2u$

[6,7]   SERROR: Isolated radical even though it does not contain the goal variable. The radical symbol appears to continue to dominate attention. Appears to be setting up another application of the square-both-sides procedure.

Since both terms in u and v still appear on both sides, she may still be attending to v as the goal variable even as she attempts to obtain an equation form without radical symbols.

Did not note the expressions are quadratic in v.

7) $-2v\sqrt{u} = 4v^2 + 3v^2u - 1$

[7,8]   "Messy-heuristic": May have anticipated that applying the square-both-sides procedure is relatively complex here and results in a pattern too "messy" to be correct based upon previous experience and interprets this to mean she has made an error. Does not BACKTRACK, however.

8) STOPPED.

FIG. 9.1.   A Student Protocol for: Solve $v\sqrt{u} = 1 + 2v\sqrt{1 + u}$ for v.

solving, the protocol illustrates her emphasis on the effects of examples and practice problems in textbooks. Students infer interesting and often erroneous "knowledge" from such practice. Such strategy errors have also been studied by others (Carry, Lewis, & Bernard, 1981; Lewis, 1981).

Learning from Examples

The assumptions discussed in the introduction have also been explored and, in some cases, modeled by cognitive scientists such as Neves (1978).[6] Neves' program, ALEX (Algebra Learning from EXamples) was designed for learning to solve linear algebraic equations such as $5x - 4 = 8x + 8$ from examples in textbooks. The program "behaves" in a manner consistent with many teachers' perceptions of their students. "The textbook provides explanatory text, annotated example problems, and work problems at the end of the section. The program uses the examples and test problems to learn and ignores the written text." (p. 11).

Neves (1981) points out several reasons why examples play such an important role in mathematics instruction.

> Examples are guaranteed to present a sequence of actions that will lead to the solution. With direct instruction there is always the possibility (and often the probability) that important information about the task will be left out of the description. Also, when the student is being told what to do the language used to communicate with him might not be language he understands. Usually procedures in textbooks are explained in their most general (abstract) form. The student cannot relate those abstract descriptions to anything concrete in his world and so cannot learn much from it. (p. 6)

This quote effectively summarizes a critical dilemma for mathematics education: the importance of making the broader context for algebraic procedures and concepts clear to the students. That requires using terms they can understand. We argue later that this can be done only after a very careful analysis of the forms of understanding that mathematics educators expect and the creation of sets of practice tasks that are carefully designed to systematically support the inferences necessary for developing those specific forms of understanding. While textbook authors' intentions in the design of lists of practice problems may be the best, the side effects—that is, the *student's* inferences—can be catastrophic. Illustrations will support that contention.

---

[6]Rissland's (1981, 1985) work on constrained example generation is also useful here. In existing curricula far too little emphasis is given to creating examples (or counter examples) having certain constraints.

Silver (1983, 1984), Bundy (1979, 1981, 1983), and Welham (1978), at the University of Edinburgh, have also done some interesting work on learning from examples using A-level equations on university-level entrance tests in Great Britain. This work may be seen as the logical extension of the work on algebra equation solving described by Schoenfeld in chapter 1. It differs from Neves' work both in its mathematical level and in its intentions. It was not done as a model for how humans solve equations. It was done as a project in artificial intelligence to study the questions: (a) How can search be controlled in domains with a large search space? and (b) How can this control information be learned? (Here the term "learned" means by the computer.)

Examples of the equations of interest (from Silver, 1984, p. 8) are:

$$\cos(x) + \cos(3x) + \cos(5x) = 0$$
$$e^{3x} - 2e^x - 3e^{-x} = 0$$
$$\log_x x + \log_x 2 = 5$$

Silver's (1984a) dissertation is an extensive PROLOG program called LP (for Learning PRESS). It uses analyses of consecutive steps in solution strategies to infer new solution methods (i.e., previously unknown to the program). It then integrates the new methods with its previous methods. It is built on PRESS [Prolog Equation solving system], a symbolic algebra equation solving program. LP and PRESS can solve approximately 80% of the equations on the difficult A-level examinations used for college admissions decisions in Great Britain. These equations are comparable with the most difficult ones found in algebra and precalculus books published in the United States.

Silver's reasons for choosing equation solving as his domain of interest were essentially the same as ours: "Equation solving has a very large search space, yet (some) humans can perform well in this domain. The large search space is necessary to test the utility of meta-level inference . . ." (Silver, 1984, p. 1). We believe that for humans, the large size of this search space helps explain why students are often able to solve virtually all the equations in the homework on a given evening, but hardly recognize some of them in a mixed list on tests.

Silver defines a learning method called "Precondition Analysis," which combines metalevel inference with concepts from the study of planning (Sacerdoti, 1977). The use of preconditions for solution methods is directly related to the planning and strategies for equation solving that will be a major theme later in this chapter. Silver's dissertation is a good illustration of one way the three assumptions discussed in the introduction have been modeled. Although his program will not be described here, the strategies and solution methods instantiated in it have fascinating implications for the forms of understanding and the skills needed by human equation solvers. In addition, he specifies them more clearly than do many textbooks.

The extensive analysis of a task as complicated as solving a mixed list of equations not previously encountered, which is required to design such computer programs, provides significantly greater insight into the specific but tacit knowledge used by skilled human solvers than most textbook authors can be expected to have. Silver's (1984) analyses yield:

- lists of the kinds of representations needed for the computer to determine patterns on which problem-solving strategies can be based
- analyses of the algebraic expressions that also determine problem-solving strategies
- lists of the relatively few methods required to solve most A-level equations, with distinctions drawn between metalevel and object level methods.

Such work has important instructional implications.

This section has focused on the theme of learning from examples, and Silver's system provides one way of doing that. But his system is also a symbolic algebra environment with implications for the mathematics curriculum.

Symbolic Algebra Environments

Most powerful symbolic algebra computing environments that teachers may be tempted to use for instruction (such as muMath) also have some glaring but interesting weaknesses. One weakness is that these environments do not have access to some very useful equation-solving methods, for example, the one described below. We discovered this while doing a careful analysis of the task of solving mixed lists of equations using methodologies from cognitive science. This analysis led us to develop a "grammar"[7] representing hypotheses about how skilled individuals solve mixed lists of equations. One of the few key strategies in the grammar involve the pattern f(EXPR1) = f(EXPR2), with f an invertible function (or, more generally, a procedure having an "inversing" procedure, such as "f is quadratic in" EXPR where the "inversing" procedure might be the application of the quadratic formula). The solution strategy is then

$$f(EXPR1) = f(EXPR2) \rightarrow f^{-1}[f(EXPR1) = f(EXPR2)].$$

---

[7]The term grammar as used here refers to a recursive description of steps by which an individual (or perhaps a computer) carries out a specified task; here, solving a mixed list of equations. The grammar we devised contains the most important equation-solving strategies.

This method is useful in solving equations such as $e^{-x} = e^{3x-1}$, for example, but it is not "known" to the versions we have of muMath, LP, or to another symbolic algebra environment designed specifically for use in mathematics instruction and soon to be published. These programs cannot solve this equation. Neither is this strategy treated systematically in textbooks. "Rewrite" strategies, which replace the equation by an equivalent one using logarithms, are more common in both textbooks and symbolic algebra environments. This is unfortunate, since thinking of equations as objects on which one operates may be conceptually more consistent with the use of symbolic algebra environments than using rewrite rules.

The availability of such powerful symbolic algebra software for microcomputers raises an important question about the objectives and content of the mathematics curriculum: Should the amount of manipulative practice in algebra be reduced because computers can now perform so many of the routine algorithms for the student? Recent reports from the National Council of Teachers of Mathematics (1984) have recommended doing just that. But this question is difficult to answer, and the mathematics education and research communities have not been sufficiently explicit about the forms of algebraic knowledge (beyond procedural levels) that they expect the curriculum and the textbooks to develop.

It may be helpful to pose the question for a more limited domain: What forms of algebraic skill or knowledge are needed to use powerful symbolic algebra environments *with understanding*? And in a logical extension of that question, we can ask: What methodologies can be used to identify those forms of algebraic understanding that will be important in the future? We should attempt to answer these questions and others like them before making major revisions in the curriculum. The next section draws on the perspectives and methodologies of cognitive science to illustrate some excesses of existing text materials and begin to clarify some forms of algebraic understanding. Currently, these forms of understanding are not effectively taught, but they will certainly be important even in the presence of powerful computer algebra environments.

A CASE STUDY: EFFECTS OF PRACTICE TASKS ON ALGEBRAIC BEHAVIOR AND SOME POSSIBLE REMEDIES

The projects discussed earlier were selected because we have found them useful in interpreting observations of the day-to-day behaviors of our students. The assumptions and their corollaries listed in the introduction also play a central role. In this section, students' behavior on Equation #1 illustrates the side effects of examples and sets of practice problems in textbooks. Studying these behaviors has allowed us to:

- develop some methodologies to help remedy these effects
- use methodologies drawn from cognitive science to identify some of the deeper forms of algebraic skill and understanding which are not effectively developed in textbooks
- develop principled methods for generating practice tasks that support the development of some of the forms of algebraic understanding desired of students.

Students' strategies for solving Equation #1 illustrate the value for mathematics teaching and learning of the hypotheses, methodologies, and perspectives drawn from cognitive science. Useful perspectives include assuming the student to be a type of "information processor" and viewing algebra as a "representation" and "search" process. Methodologies include focusing on the relationships among heuristics, recall and the structure of the learner's knowledge base; performing detailed "task analyses"; using concrete illustrations of "metacognitive" or "planning" behavior in the context of certain formal algebraic tasks that are currently emphasized in the curriculum; and using "mixed cues" and "production rule" methodologies for generating principled practice tasks for textbooks. Cognitive scientists provide significantly more detail in other chapters of this volume. Here we attempt instead to illustrate the influence of these hypotheses, perspectives, and methodologies on one mathematician/mathematics teacher's efforts to reach a more complete understanding of the student behavior in courses in algebra and elementary functions.

The Example: Protocols for Equation #1

$$\text{Solve } v\sqrt{u} = 1 + 2v\sqrt{1 + u} \text{ for } v.$$

A protocol of one high school student's work on the equation was provided in Fig. 9.1. The problem was also given to 64 freshman students in a precalculus course at the university. These students averaged more than 3 years of high school mathematics. Virtually all of them had had at least 2 years of algebra.

What strategies were used to attempt to solve this equation? What might be hypothesized about the relationship between these strategies and the nature of the inferences the students drew from the practice tasks that preceded this problem? The equation was designed to explore the effects of "mixed cues" in expressions on the recall, selection, and application of practiced procedures. In particular, this equation is actually linear in the goal variable v. But, as the protocols show, the salience of the radical symbol as a stimulus to apply the square-both-sides procedure dominates many students' behavior.

Figure 9.2 organizes the initial steps used by the students into four rough categories. We hypothesize that each of the four "strategies" can be traced to the dominance of linkages between certain cues in the patterns of the expressions and heavily practiced procedures. Very few students backtracked after discovering they had started the problem incorrectly. Of those who did, virtually none were successful. Only Strategy 2, eliminate radicals, will be discussed in more detail. We label this step a "strategy error" and are especially interested in exploring its causes.

Figure 9.3 summarizes all the approaches to achieving the subgoal of eliminating radicals from the equation. They range from the valid (a) to *literally* eliminating all radicals (c). These are essentially variations of the

Equation #1: Solve $v\sqrt{u} = 1 + 2v\sqrt{1 + u}$ for v $(N = 64)$

Strategy 1: Skilled $(N = 20)$
    Subgoal: Separate terms containing the goal variable from other terms.
    Step 1: $\rightarrow v\sqrt{u} - 2v\sqrt{1 + u} = 1$

Strategy 2: Eliminate Radicals: $(N = 25)$
    Salient Features: The radical symbols.

    e.g. True Linearity Error: $\rightarrow v^2u = 1 + 4v^2 (1 + u)$ $(N = 9)$
        Schema: $(A + B)^2 \rightarrow A^2 + B^2$ Context: equations (See Fig. 9.3 for others.)

Strategy 3: Divide the equation by the coefficient of one
        of the occurrences of v in the given equation;
        and violate the heuristic: the goal variable
        should not appear in the answer. Actually
        several different strategies were used here: $(N = 14)$

    e.g. $\rightarrow v = (1 + 2v\sqrt{1 + u})/\sqrt{u}$

Strategy 4: Canonical pattern: EXPR = 0 (or 0 = EXPR) $(N = 4)$
    $v\sqrt{u} - 1 + 2v\sqrt{1 + u} = 0$ for v

FIG. 9.2. Patterns Perceived and Strategies Selected for Equation #1

Two procedures were used. One student multiplied both sides of the equation by $\sqrt{u}$, perhaps recalling "rationalization" procedures. But the most prevalent behaviors are discussed next.

Procedure Selected: Square both sides of the equation

Procedure Implementation: Four different versions of the square-both-sides procedure were used by these 25 students; one valid and three with errors:

a) Valid: $\rightarrow v^2u = 1 + 4v\sqrt{1 + u} + 4v^2(1 + u)$ $(N = 10)$

b) True Linearity Error: $\rightarrow v^2u = 1 + 4v^2(1 + u)$ $(N = 9)$
    Schema: $(A + B)^2 \rightarrow A^2 + B^2$ Context: equations

c) Square only radical factors: $\rightarrow vu = 1 + 2v(1 + u)$ $(N = 3)$

d) Square only factors containing the goal variable or
    parameters: $\rightarrow v^2u = 1 + 2v^2(1 + u)$ $(N = 2)$

e) Square *twice,* factors involving radicals; i.e., c) fol-
    lowed by squaring factors under radicals a second
    time: $\rightarrow vu^2 = 2v(1 + u)^2$ $(N = 1)$

FIG. 9.3. Procedures Used to Implement Strategy 2: Eliminate-Radicals: (and related errors) (N = 25)

square-both-sides of the equation procedure. Students using (b) are making the kind of linearity error discussed by Matz (1982). The student who squared certain factors twice, (e), may have been interpreting instructions either by teachers or textbooks quite literally. When two radicals are present, one often must square an equation twice to eliminate them.

Before developing hypotheses about the causes of this behavior, it may be helpful to examine an annotated solution strategy by a hypothetical skilled equation solver. Of the twenty students who used Strategy 1, 18 followed these steps. Of those, 11 completed the problem correctly. The remaining 7 made errors—typically with signs.

Equation #1: Solve $v\sqrt{u} = 1 + 2v\sqrt{1 + u}$ for v   (N = 64)

Strategy 1: Skilled   (N = 20)

> *Subgoal:* Separate terms containing the goal variable from other terms.
>
> Step 1: $\rightarrow v\sqrt{u} - 2v\sqrt{1 + u} = 1$

Salient Features of the original equation that may have led to this subgoal *(hypothesis):*

Notice a) that there are two copies of goal variable *v* and that the forms $\sqrt{u}$ and $\sqrt{1+u}$ are independent (explicitly) of v and so can be "chunked" (mentally, say, as A and B respectively) and temporarily ignored; b) probably that each occurrence of v has no explicit exponent (so it is 1), although this may be noted after something like the pattern below is noted.

The features above and the related processing result in a new representation of the equation whose essence is illustrated by a pattern of roughly the following form:

$$vA = 1 + 2vB \text{ (or variations, e.g., } Av = 1 + 2Bv).$$

The "linear" features of this equation are closely linked to a sequence of steps heavily practiced in elementary algebra. This sequence involves "attraction" of the copies of v; knowledge that the linear copies can be "collected"; and, possibly at this stage, recall of the pattern $AX = B$, a precondition for the application of the isolation method $\{AX = B \rightarrow X = B/A\}$ for which "attraction" and "collection" are frequently subgoals when the structure of the equation is linear.

We suggest that such feature-checking mediates between the structure of the pattern and the recall of other patterns and relevant procedures in long-term memory. This leads to the first subgoal listed above.

Note that, if feature b were not noticed, it seems equally likely that these individuals would have selected the subgoal pattern: EXPR $= 0$, that is, $v\sqrt{u} - 2v\sqrt{1+u} - 1 = 0$ listed as Strategy 4 in Figure 2.

The behavior in a) (and its many nuances) is an extremely important form of algebraic skill and understanding and will be referred to as the Generalized Substitution Principle.

Step 2: $v(\sqrt{u} - 2\sqrt{1+u}) = 1$ "Collect" v's

using the distributive property as a "rewrite" rule. Note the top-level-structure of this equation is $AB = C$, which again makes use of the Generalized Substitution Principle with $B = \sqrt{u} - 2\sqrt{1+u}$. One of the preconditions for using isolation is that only one of A, B or C contains the goal variable. This leads to the application of the isolation method $\{AX = B \rightarrow X = B/A\}$.

Discussion of the Example

Linearity as a Salient Pattern in Algebraic Expressions.[8] For most students, the term "linear" is associated almost exclusively with finding equations of or graphing lines. Searching for and recognizing linear patterns in algebraic expressions is not often associated with the term and has little *strategic* importance to students. Nor do textbooks contain practice tasks designed to help students discern such structure in expressions or practice using it for planning strategies. Students seldom appear to make conscious use of patterns like this. On the other hand, protocols indicate that very skilled equation solvers (such as undergraduate teaching assistants) frequently note such patterns explicitly as they pause to plan a strategy. For them, "linearity," in the context of algebraic expressions, appears to be used as a property or attribute to organize, associate or recall a large variety of different objects, tasks and procedures. Such concepts or "nouns" in Davis' (1984 pp. 36–37) sense supply important and functional categories for the organization of the student's knowledge.

Students' skill in the management and use of their knowledge-base is also critical. Other essays in this volume have discussed metacognitive, or executive, behaviors, primarily in the context of general problem-solving tasks. We believe that such behaviors and their clarification are also important in fairly formal algebraic tasks such as in solving Equation #1 and certainly in solving mixed lists of equations or unusual equations such as:

$$\frac{x^2}{2500} + (\log x)^2 + \log x^{19} = \frac{1}{25}\log x^x + \frac{19x}{50} - 34$$

(Davis, 1984, p. 278)

Dominance of the Radical Symbol as a Cue. As Fig. 9.3 shows, protocols suggest that many students believe radical symbols are something to be eliminated: "I don't usually like leaving radicals in equations": "I don't like radicals;" "Radicals are bad—after the 60s;" "Radicals are always problems." The mere presence of the radical symbol triggers the desire to transform the equation by getting rid of the radical. Eliminating radicals by some version of a "square-both-sides" procedure is often routinely set as a subgoal, even if it is a poor strategy, as in Equation #1.

Automaticity and Planning. Through extensive practice, students appear to have developed some fairly consistent criteria for the patterns

---

[8]The emphasis on salient patterns in mathematics education is discussed by Knuth (1974, p. 334). He argues that attention should be paid to mathematical training in recognizing meaningful patterns in finite sums and bemoans his students inability ot perceive such patterns. Observing this inability in his students led him to introduce a course called "Concrete Mathematics" at Stanford University, where this kind of mathematics is taught.

desired in algebraic expressions. Students frequently eliminate radicals, denominators, and parentheses *before* examining the structure of the equation for cues to facilitate their search for productive strategies. Only then do they have a form which they believe might facilitate the recall of possible next steps.

The behavior of students who attempted to square both sides of Equation #1 indicates that the same action may be driven by different goals. Their objective (not necessarily conscious) may either have been to: (a) clear radicals rotely (just as they might "cross-multiply" to clear denominators), a routine they carry out (whether strategically wise or not) even *before* they actually examine the equation more systematically for patterns which might trigger recall of a strategy for solution; or b) clear radicals because they believed that is *the* strategy to be used for solving the equation.

Observations of student behavior lead us to hypothesize that those with the second goal are more likely than their peers in group (a) to lose track of the goal variable, v, in the equation and proceed to solve for u. After achieving a form that meets their criteria for patterns worthy of "scanning" for features that might trigger recall of solution strategies, some of those in group (a) may, in fact, tend to look for or notice salient patterns, such as quadraticness. They may *suspend* their use of such knowledge until an algebraic form meeting some fairly specific and consistent criteria is achieved using highly automatized procedures.

In fact, students' behavior on Equation #1 did not illustrate this hypothesis. Of the students who squared both sides of the equation, 10 did so properly and 9 made "true linearity errors" (see methods (a) & (b) in Fig. 9.3). Squaring both sides properly leads to:

$$v^2u = 1 + 4v\sqrt{1 + u} + 4v^2 (1 + u).$$

Squaring both sides while making the linearity error results in:

$$v^2u = 1 + 4v^2 + 4v^2u \text{ or}$$
$$0 = 1 + 4v^2 + 3v^2u.$$

Although 6 of these 19 students tried factoring techniques, none of them tried the quadratic formula. Of course, since they had not noticed the linearity of the original problem, they were unlikely to notice the quadraticness of the equations which resulted from squaring. Given students' inclination to factor out common factors using the distributive property, we were surprised that none of the six students who created the latter two equations collected terms and factored out a common v^2. The point of the example is simply to illustrate the difficulty of distinguishing the planned behavior of students from the rote.

One reason for that difficulty is that students, unlike skilled problem solvers, rarely allow time for a planning phase. This phase involves such activity as noting the location of the goal variable, v; examining the structure of the algebraic expressions $v\sqrt{u}$ and $1 + 2v\sqrt{1 + u}$; and attempting to classify the structure as a function of v. These types of planning activities characterize the metacogitive behavior discussed in the chapters by Schoenfeld and Silver, as well as in Brown and DeLoache (1978) and Schoenfeld (1983).

The lack of a planning phase also illustrates differences in beliefs about what doing this type of mathematical activity is about. These students do not consciously reject planning; it is just not part of their idea of doing mathematics. For them, doing mathematics means applying the first procedure that comes to mind without checking for the necessary preconditions. It also means not backtracking if the chosen procedure appears to lead nowhere.

Planning and the Generalized Substitution Principle

Consciously pausing to examine the structure of the expressions in the equation is critical for planning strategy. Having the skill to recognize that these expressions are "linear in v" *before* rotely manipulating the equation (into a form that satisfies the student's criteria for scanning) is critical for recalling the procedures required to solve it.

The ability to recognize features such as linearity requires subtle but very important forms of algebraic understanding that are not systematically developed or taught in textbooks; that is, it requires the ability to determine the features of an algebraic expression that are salient for a given task. Students must recognize that they can, in effect, *replace* by a single symbol all factors and/or terms that contain parameters other than v. Using the methodologies of cognitive science, it would be useful to perform a more complete task analysis of this skill, or rather set of skills, that we have labeled the *Generalized Substitution Principle*. The principle involves subtle forms of equivalence and requires significant algebraic skill for its application in specific contexts. Another example of the principle involves nested bracket tasks such as writing $-[x - 2(1 - x)]$ in an equivalent form without brackets or parentheses; it is discussed by Wenger and Brooks (1984, p. 229).

Explicitly identifying the kinds of tacit knowledge relevant to the principle and analyzing them are tasks for cognitive science. Its perspectives will help us think about students' mathematical behavior and completely analyze the related mathematics tasks.

Mixed Cues and Parameters in Expressions

The perspectives described here have led to an hypothesis that explains why students seem to have such difficulty with tasks—such as equation solving—involving expressions that contain literals or parameters. It is very difficult to create algebraic expressions or equations that contain mixed cues, such as linear forms and radical forms, without using parameters. Thus, if relatively few tasks involve parameters, the learner is unlikely to develop the pattern discrimination skills desired. Compare the following equations with Equation #1 for example:

$$\text{Solve } 3\sqrt{x} = 1 + 2\sqrt{x + 1} \text{ or } 3x = 1 + 2x \text{ for } x.$$

The structure of these equations is much more transparent than that of Equation #1.

If solving such equations is viewed as a search process, the combinatorics of checking for salient patterns or recognizing them increase rapidly as parameters are included. This is because so many more mixed cues must be processed. Thus, if we want students to develop pattern discrimination, we must provide extensive practice with carefully designed problems containing mixed cues and parameters.

Role of Textbook Practice Tasks

One reason students have so much difficulty recognizing patterns in equations containing mixed cues and parameters is the poorly designed sequences of exercises in textbooks.

Virtually all second-year algebra and precalculus texts contain problems of the following type:

$$\text{Solve the equation } \sqrt{x + 1} - \sqrt{x} = 2 \text{ for } x.$$

In fact, books often contain a great many such problems, usually all in one list. I doubt that the forms of understanding learned from these particular equations deserve such emphasis. In addition, as we have seen, such excessive practice has troublesome side effects. The most extreme cases we have examined is a text in which there are more than 50 such problems scattered over 200 pages (Saxon, 1981, 1983). That text contains no equations involving radicals other than this type. While scattering mixed lists of problems throughout the book is an effective teaching strategy, the problems chosen are generally homogeneous. We hypothesize that this approach to practice produces fewer procedural errors when students square both sides of an equation but that it results in the kinds of predictable and damaging patterns of rote behavior we have called "strategy errors."

Indeed, our student protocols show that extensive practice with such homogeneous sets of exercises leads to the following stimulus-response pattern *even when it is inappropriate as a strategy.*

IF [radical symbols (mainly square roots) appear]
THEN [separate-radicals and square-both-sides].

This strategy works, of course, for virtually all the problems in the list, so, barring other errors, it is reinforced. Anderson (1983) would say that it has a high "degree of match" and "production strength." We predicted that when given patterns such as those in Equation #1, many students would square both sides. This was confirmed by protocols. The radical, as a notational cue, is stronger or more easily checked than the syntactic cues that show the equation to be *linear* in the goal variable v. Pausing to scan the expressions that make up the equation and check their top level structures (here linear) is an important form of "planning" or "control" behavior that is *not* effectively or explicitly developed in most textbooks. Yet, the ability to note such structure is a critical form of understanding that cues the recall of relevant strategies and procedures for skillful equation solvers. We must, therefore, make an attempt to determine the best ways of designing practice tasks so that they will lead to the development of skillful behavior.

To do that, however, we must first identify the important features of the skilled mathematician's insight. That is the goal of the work discussed in the next section.

Analysis of Preconditions

The important forms of understanding that are used by mathematicians to categorize or transform algebraic expressions can be identified by systematically analyzing the "preconditions" that must be present before the classification process or transformation procedures are applied. In effect, preconditions determine the salient patterns in expressions and their linkages to the forms of understanding which students are expected to infer from the practice tasks.

The presence of the radical symbol alone *should* trigger *recall* of the "inversing procedure," square-both-sides. But there is a more complete and discriminating production rule listing the most salient patterns in the structure of the equation that must exist *before* the procedure is actually applied:

Given an equation EXPR1 = EXPR2:

IF   a) a radical (of index 2) appears in either EXPR1 or EXPR2

and

b) an expression containing a letter appears under at least one radical

and

c) at least one letter under a radical is the goal variable

and

d) there is no operator in the equation "above" the radical operator except $+$, $-$, $*$, or $/$

THEN   a) separate-radicals
          and
          b) square-both-sides of the equation.

This more discriminating rule illustrates a powerful methodology for generating practice tasks more systematically. A primary purpose of the practice tasks is to permit the student to infer when the target procedure or strategy might be fruitfully applied and when not. A more general (and perhaps more important) purpose is to help students learn to perceive and use relevant patterns in algebraic expressions to create strategies and recall procedures for carrying them out. Note, for example, the knowledge required to clarify the meaning of and/or check for the presence of IF condition (d). Many facets of the Generalized Substitution Principle mentioned in a previous example are implicit there.

The preconditions for such a procedure often make explicit the kinds of "understanding" that mathematics educators hope their students will infer from their practice tasks. But these kinds of understanding are seldom made sufficiently explicit to allow many students to infer them. For example, very important forms of understanding are required to check some of the conditions just listed, especially those loosely described in (d). Such "production rule" descriptions, even incomplete ones like the foregoing one, make the forms of understanding sought significantly more explicit than other types of descriptions do. This explicitness can be exploited to develop text (and computer-based) materials that support students' attempts to infer salient patterns and their use in planning strategies and recalling relevant procedures for carrying them out. A more concrete illustration of these ideas is provided in the next section.

Production Rule Methodology for Generating Practice Tasks[9]:

One method for generating practice problems that support effective student inferences is to examine the production rule, select each subset of its IF conditions and create an equation (or inequality) that satisfies, if possible, only those conditions contained in the subset. For example, applying the methodology to the rule cited systematically generates problems such as the following:

---

[9]"Production systems" are built from such rules. They are used in artificial intelligence to build expert systems and by many cognitive scientists to build computer models representing hypotheses about how learners carry out tasks. These efforts not only help to test the hypotheses but make the important forms of tacit knowledge required for the task much more explicit. See Anderson (1983, pp. 7–12) for example.

Equation 1: $\sqrt{2} + 3x = \sqrt{5}x$ (Satisfies a but not b or c)
Equation 2: $\sqrt{x+1} - \sqrt{x} = 2$ (Satisfies a, b & c)
Equation 3: $v\sqrt{u} = 1 + 2v\sqrt{1+u}$ for v. (Satisfies a & b but not c)
Equation 4: $y\sqrt{x} = 1 + 2y^2\sqrt{1+x}$ for y. (Satisfies a & b but not
                                                    c)

Note that Equations 3 and 4 are created using the methodology in combination with important patterns *previously* studied by the student, that is, linear and quadratic expressions and equations.

We have been collecting sets of examples for techniques like factoring, squaring-both-sides, grouping, and clearing the denominators. Textbook terminology, student behavior, and comments in the protocols facilitate the identification of the most prevalent linkages of notational and syntactic cues with the recall and application of procedures.

These initial efforts have made us skillful at evaluating lists of textbook exercises and at *predicting* troublesome side-effects. But, more important, as the production rule approach illustrates, these methodologies also have immediate by-products. They are the identification of the forms of knowledge and discrimination that we normally associate with algebraic understanding; the development of criteria for producing robust lists of practice tasks that facilitate such understanding; and the subsequent development of a system for applying these criteria. Many of these new tasks fit nicely into existing textbooks. Others, such as some of the "transcendent" strategies for equation solving identified in the grammar mentioned previously, require more explicit attention to features of algebraic expressions (and their meaning) than existing books provide.[10]

Some Perspectives and Criteria for the Design of Textbooks and Curricula

Figure 9.4 summarizes the relationships among several of the primary themes of this paper.

It also lists a few of the key pedagogical decisions to be made. The first, the choice of task to be analyzed, is central, because the time and

---

[10]We do *not* advocate explicitly teaching the conditions although this may be appropriate for some. Most of this analysis is to be invisible to the student, but it should be explicitly used by authors in the design of textbook exercises. The students then discover or infer the salient patterns and, we hypothesize, create their own more robust forms of discrimination and awareness of important algebraic patterns and relationships. They then use these understandings to trigger recall and test for the application of appropriate strategies and procedures at appropriate times. It appears to be important that students develop this way of thinking about algebra. And the result may be "beliefs" about mathematics that are very different from current ones. We are also experimenting with tasks to facilitate planning behaviors.

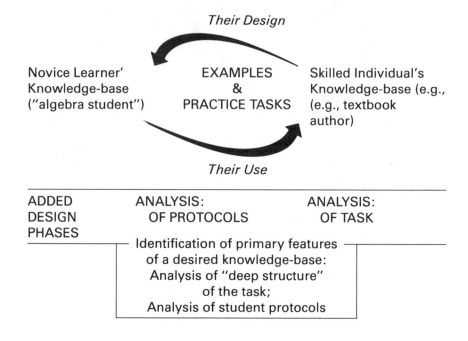

Their Design

| Novice Learner' Knowledge-base ("algebra student") | EXAMPLES & PRACTICE TASKS | Skilled Individual's Knowledge-base (e.g., (e.g., textbook author) |

Their Use

| ADDED DESIGN PHASES | ANALYSIS: OF PROTOCOLS | ANALYSIS: OF TASK |

Identification of primary features
of a desired knowledge-base:
Analysis of "deep structure"
of the task;
Analysis of student protocols

SOME KEY PEDAGOGICAL DECISIONS:

> **. . . Choice of task to be analyzed;**
> **. . . What should be explicitly practiced, made automatic?**

FIG. 9.4. Knowledge-base Paradigm for: The Metamorphosis of Novice to Skilled Performance on a Complex Mathematics Task; Implications for Curriculum Development; Implications for Design of Instructional Strategies & Materials

expertise needed to carry out all phases outlined in the figure can be significant. The criteria used to make these choices involve contemporary issues concerning the content of the curriculum in mathematics education such as the emphasis on the relationship of procedural skill to conceptual development and the identification of the effects of the computer-based symbolic algebra environments discussed earlier. Our efforts to identify the kinds of knowledge required to use these computing environments with understanding have motivated extensive analyses of the kinds of information present in the structure of algebraic expressions and students' use of such information.

The results of these analyses should be useful to authors of textbooks. As it stands now, the majority of published books for classroom use are written by skilled mathematics teachers and mathematicians and are based on their extensive experience. This is useful, as far as it goes. To be more certain that textbook problems will develop important kinds of understanding, textbook writers must identify the deeper objectives of the curriculum in algebra and elementary functions that we have alluded to and must carefully design sets of practice tasks. The variety of perspectives required for such an approach may require new kinds of textbook writing teams that represent extensive mathematical knowledge, extensive teaching experience, and perspectives of the kinds summarized here.

Many textbooks and their practice problems are essentially permutations of previous books. Consequently, new texts generally go no further than the old ones in their rationales for and designs of practice problems, and mathematics educators may remain puzzled by their students' inability to replicate *their* understanding of algebra. Furthermore, the relatively ad hoc manner in which practice tasks are currently selected and sequenced often does little to develop the many forms of algebraic understanding, for the emphasis in these tasks is nearly always quite procedural. Our discussion of solution attempts for equation #1 illustrated the results of such an approach.

Perspectives from systematic analyses of both student protocols and the deep structure of the tasks appear to be quite useful. This paper and others in this volume illustrate some of the methodologies and points of view developed by cognitive scientists to facilitate such analyses.

One last caveat: The use of equation solving to illustrate the ideas in this paper should not be misunderstood. Making people better equation solvers is *not* an appropriate goal for the mathematics curriculum during the last part of the 20th century, in my opinion. Efforts to identify the forms of skill and understanding that might be the residue of the great emphasis on equation solving are quite important, however. It is these efforts which will clarify the forms of algebraic understanding most needed; provide specific ideas about the nature of the practice tasks required to develop appropriate student knowledge; and imply criteria for the evaluation of existing instructional strategies and materials. We believe that comparable analyses should be carried out for a large number of topics currently stressed in courses and textbooks.

Candidates for the kinds of analyses proposed include topics on which the current curriculum spends a great deal of time, often several years. Figure 9.5 provides a "skeleton" for an important class of them: the forms of representation of the abstract concept, function. Each of the vectors in the diagram symbolizes a correspondence between different representations of the attributes of functions. Fluency in translating from one representation to another and heuristics for determining which representation will be most useful for a task are central to the design of effective curricula.

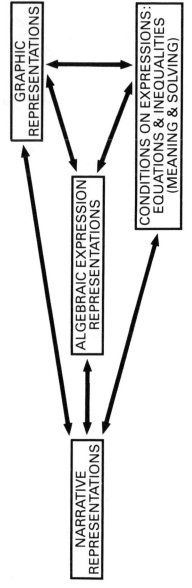

FIG. 9.5.   Representations of functions: Diads for more complete task analyses

While this paper has focused on the diad, expressions ↔ equations, one could also focus on changes of representation such as formal algebraic ↔ graphic or narrative ↔ graphic. Many textbooks provide relatively little practice in the latter, for example. More systematic identification is needed of the forms of understanding that are required to see and use the correspondence between the patterns in the structure of an algebraic expression and the features of its graphic representation.

SUMMARY

Our desire to understand the behavior of our students better motivated our interest in the cognitive science work on mathematics learning. Our purpose has been to summarize some of the work we have found influential and to illustrate the nature of that influence in concrete terms. We have begun to explore such phenomena as how strategies are planned or recalled in solving equations; the relationship of such strategies to practice tasks; the identification and use of salient patterns in algebraic expressions as heuristics for developing strategies and for changing the representations of functions; and some approaches to the systematic generation of practice tasks that develop the forms of algebraic understanding used by those skilled in mathematics.

We have made some progress in clarifying the forms of algebraic understanding desired by mathematics educators. For example, we can give more explicit descriptions than previously of the patterns in algebraic expressions that skilled mathematicians consider salient in various contexts or use to structure their knowledge. It is possible that skill in recognizing such patterns will continue to be important as individuals attempt to use symbolic algebra environments on computers in an informed way. But we will have to be specific about the forms of skill and understanding desired of our students, for vagueness inhibits the ability of text materials to support the inferences we expect students to make.

We believe that the work and perspectives of cognitive scientists and scholars of mathematics education has extraordinary potential for improving the quality of mathematics teaching and learning. Greeno (1978) a psychologist who continues to make important contributions to our knowledge of mathematics learning, has described that potential nicely:

> . . . by using the conceptual and technical advances that are now available . . . we can improve the clarity with which we formulate objectives for mathematical understanding in the curriculum, we can develop improved curricular materials and instructional methods for achieving those objectives, and we can do a better job in research in evaluating the success of

various teaching methods that relate to conceptual understanding. (p. 267) [and that] . . . it seems feasible to try to work toward a curriculum in which activities and materials designed to improve student's skills and those designed to improve their understanding can be integrated and made to be synergistic rather than competitive and antagonistic. The development and evaluation of that curriculum seems to me to be an exiciting potential task for research in mathematics education. (p. 282)

We share his optimism and hope that the content of this paper helps to illuminate ways that perspectives and methodologies from cognitive science can be used to improve the secondary mathematics curriculum and the design of instructional materials to support it.

ACKNOWLEDGMENT

I would like to thank Jane Schoenfeld and Alan Schoenfeld for their suggestions and editorial work on this chapter.

DISCUSSION

Jim Kaput. I would like to react to your statement that your work is basically informed by the buggy work. The buggy work, as I understand it, has to do with knowledge in an essentially self-contained domain. Subtraction was chosen as the domain of study because it was self-contained. To treat algebra as if it were a self-contained domain is to commit a profound error. The basic problem with learning algebra is that it is a language without semantic content. The semantic content that it does have has developed among experts through their manipulation of sophisticated algebraic expressions over long periods of time. It has a sort of *invented* semantic content. Since the more complicated algebraic maneuvers will be done increasingly by machines anyway, the semantic content of algebra for the ordinary person lies elsewhere. I think the primary source of meaning for algebraic expressions is in their capturing of numerical patterns from arithmetic, from science, or from anywhere you find them. So I think that a key problem in algebra has to do with the lack of semantic content, and that problem will be exacerbated by the development of technological instruments that will take over some of the more complex algebraic manipulations.

Ron Wenger. I don't disagree. I think we should be concerned with the fact that there are at least two critical categories of knowledge that need to be related. One is the kind of prealgebra knowledge that students have in addition to their numerical knowledge. The semantics of equations, for example, is a part of that knowledge. The other is what I'll refer to as prescriptive knowledge consisting of what we mathematics educators and mathematicians have come to view as an important way of thinking about algebraic expressions. Somehow we've got to bring those two things together. The ultimate task then is to design instructional experiences and materials that somehow bridge that gap. I don't disagree with your comment at all. It's an important one, but let's not lose the important prescriptive knowledge either.

Jim Kaput. My second point is that I think this is a good example of an application of research that is in danger of ratifying the curriculum *status quo*. I am more interested in seeing research used to design and inspect potential future curricula. In this case, the things that students are doing in algebra classes may be more a function of curricular history than of current need.

I see algebra as one of the two principal representational tools that we try to teach students. The other has been base-10 arithmetic. In the long run, there is going to be a broadening of the collection of representational tools available to people that includes traditional algebra as but one representational tool among a variety of options. Algebra, as it was perceived by Euler and his mathematical contemporaries, only represents the world as it is. New forms of representation may be useful for representing a wider span of human experience. I'm thinking, for example, of the new versions of Logo. I want to keep a kind of broad historical perspective in the discussion because these alternative forms of representation are relatively recent phenomena. I think 20 years from now we will be using a variety of representational tools whose capacity will dwarf that of the limited ones that we have at present. I'm worried that research is being used to ratify the *status quo* and impede the development of potentially interesting and valuable future curricula.

Ed Silver. I would like to make a couple of quick points. I think Jim has raised an interesting point about research on algebra learning and understanding. One of the things that makes this a particularly interesting area right now is that it's at the center of this tremendous storm. The winds of change are whipping around the outskirts of the algebra curriculum for a variety of reasons. Some concerns are generated by research findings about what people do and don't learn very well. Also there is pressure from within mathematics and computer science to emphasize discrete mathematical topics as opposed to more continuous mathematics and the study of calculus. A curriculum that has been with us for quite some time now is in a state of flux and under close examination. It is not clear where to place your bets. As Jim points out, what much of Ron has described is an application of research that certainly can be useful in teaching the present curriculum and analyzing the present curriculum in much more detail. Alternatively, if you place your bets on a different algebra curriculum, then that sort of application doesn't appear as important and interesting. If you look at the work that Ron described globally, there is no reason why that same approach could not also be applied in task domains that have not received so much attention.

Ron indicated that he hadn't planned to talk generally about algebra, but I think we should at least acknowledge that there is a fairly substantial research base that exists on some other algebraic topics, some of which are of great concern now as educators reanalyze the pre-algebra curriculum. Those are the notions of variable and equation. Algebra, treated as generalized arithmetic, carries with it certain ways of building on students' numerical experience. That approach is also vulnerable to the deficits in the arithmetic experience that students bring with them. Work that Carolyn Keiran and Nick Herscovics have done on students' understanding an equation suggests that arithmetic understanding of equations is different from the algebraic understanding of equations needed to perform well in algebra courses. I think that kind of research points to different curricular implications and further work that needs to be done. Sigrid Wagner's work, and that of other people in Massachusetts, points to basic misunderstandings of the notion of variable. This notion is crucially important and needs careful development. So, there is a substantial research base and, if there were a consensus on the goals of the algebra curriculum, we would be in a good position to apply that research base. It is this shaky foundation of goals that makes it difficult.

One of the interesting things about taking these various pieces of research and attempting to build a new algebra curriculum is that most of the topics have been researched in relative isolation. You can gain some very nice insights into notions of equations, for example, if that's what you concentrate on. You can build up some insights on the notion of variable if that is what you concentrate on. You can build up some insights into other areas if you concentrate on those areas. What may become a problem is that in furthering the understanding of equation you might, in fact, inhibit better understanding of variable. That is not a trivial problem but one that needs attention before these ideas begin to be applied in the curriculum.

Jim Greeno.   I'd like to express my admiration for Ron's research. It's a strikingly good example of subject-matter oriented research that makes critical use of resources that have been developed in other domains to solve tough problems in the design of testing and in the teaching of algebra. It shows also that the use of those concepts doesn't come easily. The fact that somebody developed the notion of a production system doesn't automatically produce the tool with which to design tests or to understand the specific components of the curriculum. There's additional applied research required to build that bridge. It seems to me that a major contribution of Ron's work is to build some of those bridges.

My other comment is a substantive one about the cognition of algebra. I'll make a brief background comment first. I think we could make an argument that arithmetic is only about counting. We now have some understanding of preschoolers' cognitions about number before they start learning arithmetic, and they have a pretty good conceptual grasp of what numbers are in the context of counting things. This means that there's a nice foundation for beginning to learn arithmetic. But there's a fundamental shift in cognition about number from the time when youngsters only know how to count to the time when numbers become conceptual entities, things that youngsters can think about themselves. Then they learn a whole conceptual structure in which they know properties of numbers and relations between numbers and operations on numbers.

I think there may be some similar development in the shift from arithmetic to algebra. What might that shift entail? When you know arithmetic, you know many functions and you can evaluate them. But you don't have a conceptual structure that defines functions, the relationships between them, or how to operate on them. It may be that an equally fundamental conceptual shift has to occur, and algebra is a structure at another level of abstraction. We have the beginnings of a model of that as we already recognize that you have to understand the concept of the variable and that an equation is a relation rather than an operation. I think we need a clearer conception of what kind of cognitive structures are involved. I'd like to say that I have some answers but that isn't the case.

Ron Wenger. Let me make one more comment in self-defense. If this work seems related only to the *status quo,* I have not made my purpose clear. If one asks what modes of algebraic understanding you need to use a powerful computer software utility with understanding, the answer is that we don't know. We mathematicians think we know what these forms of algebraic understanding are. If there is a main point to what I was saying, it is that we don't. We can't be explicit about what they are. So I don't object at all to continuing to use our instincts for developing interesting computer environments. We are doing some of that ourselves. But to take the position that a new algebra curriculum or new ways of thinking about these kinds of tools doesn't require a much greater understanding of what we have been doing for the past two hundred years is a mistake. We need to do much more analysis instead of just flying by the seat of our pants. I don't want to throw out big chunks of the curriculum because we think symbolic algebra packages will solve our equations for us. The development of such packages may be inevitable, but I think we need a much better understanding of what we are currently doing.

Steve Maurer.   I think the most interesting point Ron has raised is about what mathematicians expect *now*. When we ask ourselves that, we ask what are we trying to teach; and we are always trying to teach general principles. I would say that I want students to be able to transform one algebraic expression into another that is more useful in that context. I want them to understand that certain algebraic operations give you equivalent statements that lead to solution sets and some irreversibly change the solution sets. Those basic principles are the heart of the curriculum.

Inspired by what Ron is doing, I realize that I should look very carefully at how I myself use algebra. One of the specific things that I want to do is look at all my mathematics research papers and examine where I used algebra in them. Because I already know some of the examples, I suspect that I would find a fairly subtle ability to do algebraic manipulations was necessary. If I looked also at books I have written and the tests that I write, I suspect I would find they require more of ability to do manipulations than I would care to admit. You can say that won't be the case for the future. You may be right, but I have experimented with muMath, and it is very hard to use those systems to do more than solve certain kinds of equations. That is not an important issue to me; rather it is a question of transforming things from one form to the other. You can use those systems to transform expressions from one form to another in some cases, but you have to be clear and persistent about what you want to do. Now there may be some future system that will do it for you easily, but if several different representations are used, it's going to be hard to get the system to know which one is correct. Somewhere in the collective consciousness must be retained what is the intuitively plausible algebraic representation for a particular problem. Perhaps students can learn without a great deal of specific pencil and paper work. Perhaps they can learn by seeing a lot of examples solved for them by a computer, but it's not clear to me how. In summary, I hope to go home and do this work. We need to come to a better idea of what we are asking of students. With that information in hand, we can go ahead and start thinking about what it means to understand algebraic expressions.

Warren Page.   I agree with Ron's agenda. Because I am a mathematician, I appreciate his point about algebra as a spiraling, hierarchical discipline composed of relations on the relations on the relations in ascending order. When Jim talks about learning representations, I get a little bit anxious. There are trade offs, trade offs in the ability to generalize and extend to higher levels of thinking. Earlier Jim Kaput gave the example of how spreadsheet programs can be used to solve maximum/ minimum calculus problems. It's true that that's an acceptable representation with which to solve those problems. The question for me is what one gains by using that representation, other than having established an ad hoc alternative process for solving a particular type of problem. The key issue is the relationship among representations. Do we simply build different representations for different topics? How are the representations related? Are some better than others for certain purposes? Is there some coherency underlying the diversity?

REFERENCES

Anderson, J. R. (1983). *The architecture of cognition*. Cambridge, MA: Harvard University Press.

Booth, L. R. (1984). *Algebra: Children's strategies and errors: A report of the strategies and errors in secondary mathematics project*. Windsor, Eng.: Neer Nelson.

Brown, A. L., & DeLoache, J. S. (1978). Skills, plans and self-regulation. In R. S. Siegler (Ed.), *Children's thinking: What develops* (pp. 3–35). Hillsdale, NJ: Lawrence Erlbaum Associates.

Brown, J. S., & Burton, R. B. (1978). Diagnostic models for procedural bugs in basic mathematical skills. *Cognitive Science, 2,* 155–192.

Bundy, A. (1979). A treatise on elementary equation solving. (DAI Working Paper No. 51). University of Edinburgh, Department of Artificial Intelligence.

Bundy, A. (1983). *The computer modelling of mathematical reasoning*. London: Academic Press.

Bundy, A., & Sterling, L. S. (1981). Meta-level inference in algebra. (Research Paper 164). University of Edinburgh, Department of Artificial Intelligence.

Carry, L. R., Lewis, C., & Bernard, J. E. (1981). *Psychology of equation solving: An information processing study*. University of Texas at Austin, Department of Curriculum & Instruction.

Clement, J., Lochhead, J., & Soloway, E. (1980). Positive effects of computer programming on students' understanding of variables and equations. *Proceedings of the National ACM Conference,* Nashville, TN.

Davis, R. B. (1984). *Learning mathematics: The cognitive science approach to mathematics education*. London: Croom Helm.

Gentner, D., & Stevens, A. L. (1983). *Mental models*. Hillsdale, NJ: Lawrence Erlbaum Associates.

Greeno, J. G. (1978). Understanding and procedural knowledge in mathematics education. *Educational Psychologist, 12,* 262–283.

Heid, M. K. (1983). Calculus with muMath: Implications for curriculum reform. *The Computing Teacher,* Nov. 6, p. 46–49.

Kieran, C. (1979a). Children's operational thinking within the context of bracketing and the order of operations, *Proceedings of the 3rd International Conference for the Psychology of Mathematics Education,* Warwick, Eng. (pp. 128–132).

Kieran, C. (1979b). Constructing meaning for first-degree equations in one unknown, *Proceedings of the 3rd International Conference for the Psychology of Mathematics Education*, Warwick, Eng. (pp. 133–134).

Kieran, C. (1981). Concepts associated with the equality symbol, *Educational Studies in Mathematics, 12*, 317–326.

Knuth, D. E. (1974). Computer science and its relation to mathematics, *Mathematical Monthly, 81*, 323–343.

Lewis, C. (1981). Skill in algebra. In John R. Anderson (Ed.), *Cognitive skills and their acquisition*, (pp. 85–110). Hillsdale, NJ: Lawrence Erlbaum Associates.

Lochhead, J., & Clement, J. (Eds.) (1979). *Cognitive process instruction*. Philadelphia: Franklin Institute Press.

Matz, M. (1980). *Towards a computational model of algebraic competence*. Unpublished master's thesis, MIT.

Matz, M. (1982). Towards a process model for high school algebra errors. In D. Sleeman & J. S. Brown (Eds.), *Intelligent Tutoring Systems* (pp. 25–50). London: Academic Press.

National Council of Teachers of Mathematics (1984). *The impact of computing technology on school mathematics*. Reston, VA: Author.

Neves, D. M. (1978). *Learning procedures from examples*. Unpublished doctoral dissertation, Carnegie-Mellon University.

Reif, F. (1983). *Understanding and teaching problem solving in physics*. Lecture at the International Summer Workshop, Research on Physics Education, La Londe les Maures, France.

Resnick, L., & Ford, W. W. (1981). *Psychology of mathematics for instruction*. Hillsdale, NJ: Lawrence Erlbaum Associates.

Rissland, E. L. (1981). Constrained example generation (Tech. Rep. 81-24). Amherst: University of Massachusetts.

Rissland, E. L. (1985). Artificial intelligence and the learning of mathematics: A tutorial sampling. In E. A. Silver (Ed.), *Teaching and learning mathematical problem solving: Multiple research perspectives* (pp. 147–176). Hillsdale, NJ: Lawrence Erlbaum Associates.

Sacerdoti, E. D. (1977). *A structure for plans and behavior*. New York: Elsevier.

Saxon, J. H., Jr. (1981). *Algebra I: An incremental development. Norman, OK: Grassdale*.

Saxon, J. H., Jr. (1983). *Algebra 1½: An incremental development*. Norman, OK: Grassdale.

Schoenfeld, A. H. (1983). Episodes and executive decisions in mathematical problem solving. In R. Lesh & M. Landau (Eds.), *Acquisition of mathematics concepts and processes* (pp. 345–395). New York: Academic Press.

Schoenfeld, A. H. (1985). *Mathematical problem solving*. New York: Academic Press.

Silver, B. (1983). Learning equation solving methods from worked examples *Proceedings of the International Machine Learning Workshop* (pp. 99–104). University of Illinois, June, 1983. Also available from the Department of Artificial Intelligence, University of Edinburgh, Research Paper 188.

Silver, B. (1984a). Using meta-level inference to constrain search and to learn strategies in equation solving. Unpublished doctoral dissertation. University of Edinburgh, Department of Artificial Intelligence.

Silver, B. (1984b). Using meta-level inference to constrain search and to learn strategies in equation solving. Revised form of doctoral dissertation. University of Edinburgh, Scotland, Department of Artificial Intelligence.

Silver, E. A. (1981). Recall of mathematical problem information: Solving related problems. *Journal of Research in Mathematics Education, 12*, 54–64.

Silver, E. A. (1982). Knowledge organization and mathematical problem solving. In F.

Lester & J. Garofalo (Eds.), *Mathematical problem solving: Issues in research* (pp. 15–25).

Sleeman, D. H., & Brown, J. S. (1982). *Intelligent tutoring system*. London: Academic Press.

VanLehn, K. (1981). *Bugs are not enough: Empirical studies of bugs impasses and repairs in procedural skills*. Cognitive and Instructional Sciences Series Report CIS-11. Palo Alto, CA: Xerox Palo Alto Research Center.

VanLehn, K. (1983). *Felicity conditions for human skills acquisition: Validating an AI-based theory*. Cognitive and Instructional Sciences Series Report CIS-21. Palo Alto, CA: Xerox Palo Alto Research Center.

Wagner, S., Rachlin, S. L., & Jensen, R. J. (1984). *Algebra learning project: Final report*. (Rep. on NIE Proj. 400-81-0028). University of Georgia, Athens, Georgia.

Welham, B., & Bundy, A. (1978). Equation solving: A progress report. (DAI Working Paper 29). University of Edinburgh, Department of Artificial Intelligence.

Wenger, R., & Brooks, M. (1984). Diagnostic uses of computers in precalculus mathematics and their implications for instruction. In V. P. Hansen (Ed.), *Computers in mathematics education* (pp. 217–231). Reston, VA: National Council of Teachers of Mathematics.

10 Cognitive Science and Mathematics Education: A Mathematician's Perspective

H. O. Pollak
Bell Communications Research
Morristown, NJ

This volume, and the conference from which it stems, have as their theme the connection among mathematics, mathematics education, and research work in cognitive science, and I, for one, am delighted. Frequently, when mathematicians hear about artificial intelligence, information technology, and cognitive science, these connections are not made. Mathematicians have to ask people to remember that there's a subject-matter out there and that the important question is the impact of these new areas on teaching that subject matter. All too often, there is a tendency to think of the research in artificial intelligence as a great advance by itself. It is, but educators and mathematicians also want to see it applied to education, and many researchers in AI stop short of that. Mathematicians like me must seem to be playing a broken record: What does this piece of research, this alternative model, this new technology, say about how to teach a specific mathematical topic?

Besides mathematics, mathematics education, and technology, there is still another component that concerns me, industry. My perspective is that of a nonteaching mathematician who has spent many years applying, and helping others to apply, mathematics to the real world. From that point of view, some of the discussions of problems at the meeting are not wrong-headed in any sense, but they don't go back far enough. For example, Ron Wenger's research on equations with radicals shows that students tend to square, regardless of what they are told they are solving for. They have been conditioned to do that. In real life, however, no one tells them what variable to solve for; and that's one of the things they have to figure out. The practice of mathematics in industry may lead to messy equations, and your job is to draw insights from those equations. One piece of the job may be to figure out what variable, or combination of variables, to solve for. This is part of the openendedness of mathematics, a topic worthy of much discussion.

Here is another example. In some calculus courses students see the solution for equations for planetary motion; these initially contain five variables. An early part of the job is to reduce the system somehow to two variables, so there can be a chance to solve it. There are many ways of doing that, and the textbook carefully leads the student through the one that works. It is like a well-defined path that you take through the mountains—you can't possibly get lost. You get to a well stocked hut at the end of the day, and you start off again the next day in the same way. But in the world of industrial mathematics, there may be no textbook to map the path. Education should prepare students for this possibility. I favor occasional cross-country mathematics. I would like to have students try some of the other combinations of variables to see that they don't lead anywhere. How else will they appreciate the cleverness of the actual solution?

Similarly, when you discuss the many strategies for solving problems, you identify and systematize the different possible lines of thought tried by students (see, for example, Schoenfeld's chapter 8). However, much of the work on metacognition done so far is unrealistic in at least one respect: Because the problem has come up in school, the students know that it can be solved. Somewhere in the collection of methods, most likely in the most recent chapter of the course, there is a method for doing the problem. If they keep at it long enough they will find it. At the conference, for example, Ron Wenger showed us a list of typical problems students are expected to be able to solve. I was struck by one of them: Solve: $\sin(4A - 5) \cos^2(A + 2) = 0$. I looked at the question and said, "Hey wait a minute. That's too much of a mess; it's going to be difficult!" I've built up the instinct over time that trigonometric expressions with "4As" and "As" in them together are not going to lead to quadratic equations. So I wondered how in the world I was going to do it, until I remembered the context of the problem. Then I said to myself, "Oh, wait a minute. The problem's in here. A solution has to be possible." Only then did I notice the zero on the right hand side and say "Aha! That's a very different problem! I'm just supposed to say that if the product is zero, one or the other factor is zero. I don't have to think about all that other stuff." The fact that the problem is in the book makes a lot of difference in the selection of metacognitive strategies.

If a test contains material covered much earlier in the year, or even before, students complain of unfairness. Compare this situation with the use of mathematics in the real world. When a new problem arises, you don't know which page it is on. You don't even know whether it is in the textbook! I wish we knew more about how to prepare people for that reality. We concentrate on preparing students to solve problems they *should* know how to solve. But in the real world we constantly come across new problems, which may or may not be solvable. Here is another example of taking the analysis one step further back. When people apply mathematics in engineering, for example, they may have to decide whether or not they can solve a particular equation. They make various choices depending upon that decision—should they consider a simpler model, a simulation, a discrete version, a numerical solution? What can we teach students that will help them in this process? That question precedes the question of choosing among various methodologies once you know that you can solve the problem.

We now have seen two ways in which mathematics in the real world is different from much classroom experience. A third way in which real-world mathematics is different is in the sphere of problem formulation. Jeremy Kilpatrick has examined this area beautifully in chapter 4, and I should like to underline some of his ideas. As Kilpatrick says, problem formulation in the real world has many different meanings. In my job, problem formulation means to take a fuzzy, ill-defined situation in some other field, or in the real world, for which there is no obvious mathematical formulation or structure, and formulate a mathematical problem that will help make the situation precise and quantitatively, structurally and analytically comprehensible. An example of this type of formulation question is at the end of this chapter.

We have briefly examined three special aspects of mathematics as applied in industry. Let me now continue from the point of view of a mathematician more generally than of an industrial mathematician. This conference has heard a number of discussions of the difference between arithmetic and algebra—see, for instance, Greeno's chapter 5. One difference between arithmetic and algebra that particularly interests me is that in arithmetic the numbers are all alike from the point of view of the elementary operations. It's too bad if arithmetic is taught only that way. It doesn't have to be, because there is already in arithmetic the opportunity for open-endedness, for seeing that numbers are very far from all alike. Furthermore, the difference between varying degrees of success and partial failure in algebra is very much a question of arithmetic.

Children frequently have trouble learning the multiplication table, particularly the multiplication factors with the largest answers. Which of these products gives 48; which, 54; and so on? For some children, these facts are hard to remember. A good way to help students learn them is to work backwards. "Give me a multiplication fact that gives you 14." "Thank you. Give me a multiplication fact that gives you 15." "Fine.

Give me a multiplication fact that gives you 16." "Give me another one."
"Give me a multiplication fact that gives you 17." (Pause) "What do you
mean you can't? Give me one." "Give me one that gives you 18. Give me
another one." What's happening here? This is a strange pattern in
arithmetic! Some numbers come up twice; some numbers come up once;
some numbers don't come up at all. Numbers are not all alike! This is
good; it is open-ended; and it sows the seeds for much interesting
mathematics. Furthermore, the difference among numbers multiplicatively form one of the fundamental insights behind success in algebra.
One of the things you have to know in simplifying all of those horrible
expressions and dealing with all of those equations is the biographies of
the common numbers. You have to know that when 24 shows up you must
be careful: There are many ways to get 24. Twenty-three, on the other
hand, will almost never come up; and if it does, there is only one way that
that could happen. Knowing the multiplicative biographies of the integers
will lead to success in many aspects of algebra.

An issue on which additional research would be very welcome are the
differences among various approaches to the same topic. What can
research tell us about the kind of approach that will succeed with certain
kinds of students? Greeno's chapter on representations contains some
excellent beginnings. For example, at some time we must approach the
concept of an absolute value. Offhand, I know at least five different ways
of defining $|x|$ that people have used. One definition says $|x| = x$ if x is
positive and $-x$ if x is negative. Another: $|x| = \sqrt{x^2}$. This definition
requires students to remember what square root means. $|x|$ can also mean
the nonnegative between x and $-x$. It can mean the distance from zero to
x, wherever x is on the number line. It can mean the larger of x and the
opposite of x, that is, the one to the right on the number line. Those are
five different definitions. What can research tell me about the effectiveness of each one in instruction? That's a question whose answer I need
desperately, because absolute value is a tough concept to teach.

Similarly, what rationale can we give students to explain why the
product of two negative numbers is positive? I have seen four different
approaches in texts. One approach plots points on the line $y = 2x$. You
get $2 \times 3, 2 \times 2, 2 \times 1, 2 \times 0$ in the first quadrant and note that the
sequence of answers ought to continue as a linear relationship into the
third quadrant. If it does, then you have motivated that positive times
negative is negative. You now look at the line $y = -2x$. You have already
motivated—with commutativity—that $(-2)(3) = -6, (-2)(2) = -4$. If
you continue on the line into the second quadrant, you can motivate that
the product of two negatives is positive.

Morris Kline, in a lecture at Allerton House many years ago, suggested
that one should make a movie of water being pumped into a tank, and of

water being pumped out of a tank, and then run both movies backwards. If you run the movie of the water being pumped out of a tank backwards, it looks as if water were being pumped into the tank and, "therefore" (he suggested) the product of two negative numbers is positive.

Another method is to use balance boards. Let's say you've got a balance board with a mass of 1 at 6 and a mass of 2 at -3. The board balances. Rather than saying that the sum of weight times distance on one side equals weight times distance on the other side (which the physicist has to undo later on), you can start talking about the sum of the moments from the beginning. You can see that 1 at 6 and 2 at -3 balance. Having done various examples like that, you lead up to the mathematical formulation $1(6) + 2(-3) = 0$. Next you can put a mass of 1 at -6 and a mass of 2 at -3. Of course, the system won't balance. Now you say, "Let me show you what to do. I'm going to put a pulley above -3 and run a string up and over the pulley and attach the unit mass to the string. Now the force acts in the opposite direction. Look! It balances!" If you are willing to believe that the force acting in the opposite direction ought to be negative, then indeed $(-2)(-3)$ balances 1 at -6. So $(-2)(-3) + (1)(-6)$ ought to be 0. If this pattern is to persist, it must be that $(-2)(-3) = 6$.

A fourth method uses the persistence of the distributive law and the properties of zero. You know that by definition $(3) + (-3) = 0$. If you multiply that whole equation by 2, $(2)(3) + (2)(-3)$ must be 0. Well, $2 \times 3 = 6$, so $(2)(-3)$ has to be -6. Now you do the same thing with the negative numbers. Multiply the equation, $-2 + 2 = 0$, by -3. $(-3)(2)$ is already known to be -6, so $(-3)(-2)$, if the expression is going to add up to zero, has to be 6. Whatever multiplication is like for negative numbers, it has to continue to have the distributive property. That can also be used as a rationale.

I'd like to know which, if any, of those four methods works with what students. This information would be valuable for the classroom teacher— provided, as Anna Henderson points out in chapter 6, the teacher has time to use it!

Another question: Is it better to introduce negative numbers as new objects, as signed numbers distinguished from unsigned numbers, or is it better to think of them as a geometric extension of the existing numbers? We could give names to the other half of the number line and not bother to create new entities that would then have to be identified. Do we know which approach would be pedagogically better? There are many aspects of the mathematics curriculum with different but equally valid approaches. The crucial issue is which one works best with what kind of students. As a mathematician, I want to know.

In the early 1960s, there was a lot of interest in making mathematical films. The MAA, for example, had a Committee on Educational Media, and NSF support, for this purpose. Some delightful films were made, but there was very little in them that couldn't have been done with a piece of chalk at a blackboard. What was their purpose? My point of view on Educational Media is a little different. I would like to know the toughest pedagogical problems we have and ask what new technology can do to help me teach them. At the height of the films' popularity, I wanted to know what they could help me teach. Now I want to know how some of the information technologies, and the new insights into how students learn, can help me teach the difficult topics. Roy Pea's chapter provides a good structure for thinking about this question, though it also does more: It points to interesting ways in which cognitive technologies might be used to develop and foster mathematical thinking in general.

Let us come to an issue that Joe Crosswhite raises in his paper and that was discussed in the conference session on "teacher's concerns." How do we get more expository papers of high quality? First of all, I'll give you my totally irreverent version of the answer. We don't see expository papers because there's no advantage for young professors to write them. Will the college promote you for writing expository papers? Probably not: Most schools will say you have to do research and publish the papers in refereed journals. Your colleagues have to write to the dean that this is great research, and then they'll consider promoting you. So let's forget about promotion. Will you at least get the plaudits of your colleagues? No, they will criticize you for not saying everything quite precisely, for not giving the exact conditions for some results, and for seeming to be unaware of some other, related piece of research. By writing a good expository paper, you may get nothing but trouble. You spend a lot of time on it, but you are not going to get a promotion or a raise for it, and your colleagues will jump all over you for being imprecise. What's the result? Good expository papers are written by full professors—let's be grateful!—but seldom by other people. I hope the conception of scholarly activity will broaden and that the academic community will recognize how valuable these papers really are.

Exposition in mathematics education presents another interesting problem. In mathematics, I am used to theorems. Are there theorems in mathematics education? Tell me something that is established as fact. Joe Crosswhite is quite right in encouraging us not to wait until the facts are established, but I want to know if they ever really are. Are there results that you can say are absolutely true and that future research can build upon? Cognitive science researchers hope to achieve a higher degree of certainty than traditional educational research ever could. In some sense, as Alan Schoenfeld points out, an AI program is proof that a particular outcome is achievable under particular conditions. When mathematicians go to an international conference on mathematics education, one of the natural questions they ask themselves is, "What did we hear at this conference that we couldn't have heard 4 years ago?" At conferences about mathematical research, it is very unlikely that you will hear a paper that you could have heard 4 years ago, because the bulk of the presentations are about research done in the last 4 years. At a conference on mathematics education, how much really new is there? The answer is that there is much more than there used to be, and the main reason is the subject of this meeting. It is precisely the learning that goes with information technologies, with artificial intelligence, with cognitive science, and with the inspiration from the technology. Much of this understanding wasn't there 4 years ago, and there is consequently exciting new life in this discipline. Mathematics education looks a lot better now to a mathematician who tends to look down his nose at anything except mathematics.

Ed Begle was the first to point out to me that mathematicians think that they know everything about mathematics education and that anything they instinctively believe must be fact. On the contrary, most of their beliefs are unproved—and many of them are untrue. That may be part of being a research mathematician. Along with expository writing on mathematics for the benefit of teachers, we very much need expositions that outline what we are beginning to know about mathematics learning for the benefit of professional mathematicians. Steve Maurer said this, too, in the teachers' session. These kinds of expositions would be a valuable contribution to the process of building cooperation between mathematicians and mathematics educators.

I was very interested to hear Joe Crosswhite say that research in mathematics education exists at the elementary, secondary and early college levels. Personally, I've had a difficult time finding any research in mathematics education at the college level. I can count on the fingers of one hand the number of people I know who work in this area. The scarcity is surprising, because both the available data and the available set of questions is very rich at the college level. For example, one of the perennially unsolvable problems in college mathematics is how to teach calculus. No one agrees with anyone else about how to teach it. That's why there are a hundred new texts each year on the subject. People don't even agree with themselves 10 years later, which is why they write a second book!

The problems of teaching calculus are extraordinarily difficult. There are "n" different topics, each of which must be treated before any of the others. You can't do that when "n" is much greater than 1. There is also the problem of how much rigor to build into the course. How, as a mathematician, do I hold up my head among my colleagues and resist the pressure of the engineers down the hall to teach cookbook procedures? Is there anything in the research that will help me out? What can research tell me about one approach to calculus being better than another? What can research tell me about such perennial campus arguments as large lecture sections versus small recitations? What difference do those teaching milieux make in students' learning—and in the likelihood of their electing another math course? I could use this information, because the economic arguments at the college level about class size come thick and fast. There are many other questions I need answered, and I wonder why there is so little helpful research.

I will admit that when I was president of the Mathematical Association of America (MAA) 10 years ago, one of my most obvious failures was not creating an interest in systematic education research at the college level. No one would touch it. Why not? The only explanation that I've ever received is that research on elementary and secondary teaching is done by university people. Hence, research on university teaching would have to be done by someone higher up the education ladder. Since no one is higher up than the university people, there's no one to do the necessary research!

In elementary education, the subject of estimation has become especially interesting. It has always been considered part of the curriculum but has not been taught particularly well. Calculators and computers have increased its importance because the best protection against errors in using the technology is an estimate of the expected answer. Most recent research that I have seen on the teaching of estimation has been done by Robert Reys and his associates (Reys & Begsten, 1981; Reys, Rybolt, Begsten, & Wyatt, 1982). Reys distinguishes computational estimation from measurement estimation. In the case of computational estimation, one looks for an approximate answer to a computational problem, an answer obtained readily by mental arithmetic. In measurement estimation—perhaps generalized from its usual definition—one wants an order-of-magnitude estimate for some real-world question. A favorite of the physicists is "How many piano tuners are there in Milwaukee (or San Francisco, depending on the physicist)? How many tailors are there in Hong Kong? (I remember constructing an estimate, based on the Yellow Pages in the phone book, of about 3,000.)

How many saguarro cacti more than 6 feet high are in the state of Arizona? I read that the saguarro is an endangered species. Developers tear them down when they put up new condominiums. So when I visited Arizona 2 or 3 years ago I decided to try an estimate. I came up with 10^8. Let me tell you how I arrived at that answer. In the areas where they appear, saguarros seem to be fairly regularly spaced, approximately 50 feet apart. That approximation gave me 10^2 to a linear mile, which implied 10^4 in each square mile. The region where the saguarros grow is at least 50 by 200 miles. I therefore multiplied $10^4 \times 10^4$ to arrive at my final answer. I asked a group of teachers in Arizona for their estimate, and they were at a loss as to how to begin. I don't know how close my answer is, but it is an example of measurement estimation.

A second research area of particular interest to me was brought up by Ron Wenger, namely the precision of instructions to communicate what we want students to do. What is the role of vagueness in teaching mathematics? We all know that every year a few students understand what you want by what almost seems to be the process of osmosis. But for most of them, osmosis won't work and they finally give up mathematics in part because it doesn't make sense to them. Many times we are imprecise in our statements of the objective. For example, consider factoring. If we ask students to factor "$x^2 - 4$," they know perfectly well what they are supposed to do. Suppose you ask them to factor "$x^2 + 4$." They will tell you that it cannot be done, that one can't factor the sum of two squares, only the difference of two squares. If you explain that the answer you want is "$(x + 2i)(x - 2i)$," they will complain that they did not know that this was what you meant. It's difficult to be precise about what you want when you say "factor," and the difficulty worsens. Suppose you give another sum of two squares, "$x^4 + 4$." Again, this seems to imply complex numbers, but this time you can avoid that issue entirely. You can rearrange "$x^4 + 4$" to be "$x^4 + 4x^2 + 4 - 4x^2$," the first three terms being $(x^2 + 2)^2$ and the last term "$(2x)^2$." So you have the difference of two squares, which can be factored in the usual way.

Typically the textbook will just say "factor." Students are expected to do the first problem and to say that they can't do the second problem; the textbook won't even bring up the third problem. Exactly what we do mean by "factor" is difficult to define precisely. One of the things we tried to do in the SMSG Curriculum 25 years ago was to give precise instructions to the students. I'm wondering as a result of the discussion here whether we were right. Should we leave the instructions vague and let students struggle with them? Should we let them catch on for themselves that "factor" sometimes has one meaning and other times another? I don't think we ever considered the possibility that this approach might be better than telling them, even though—or perhaps because—it's imprecise.

The conference also discussed algebraic symbolism and the unspoken inference that if there is an x, you must solve for it. I must tell you about an experiment I did on myself many years ago. It didn't ruin me for mathematics permanently, but it did ruin me for 3 days. I was taking a graduate real variables course, and we were doing a proof on uniform convergence of families of functions. I decided to experiment on myself by using ε to represent a large number, and N to represent a small one. I wanted to see what would happen. I found myself totally unable to think; I went completely to pieces and it took me several days to get over it. My intuition had been that ε was something very small and N was something very big. Not only was it ingrained, but it had become essential to my way of thinking and trying to change it was a disaster. I was very surprised. I never tried it again, and I advise you not to either, because it is very uncomfortable.

In discussing the function of technology as it cuts across the mathematics curriculum Roy Pea's paper (chapter 4) reminds me of simulation. Simulation cuts across many areas of mathematics and particularly applied mathematics. I would be very interested in finding out how one knows when to do a simulation; what distinguishes a good simulation from a poor one; and, generally, what students learn from simulations.

Anna Henderson (chapter 6) raised, and Joe Crosswhite (chapter 11) echoes the question of getting research into readable form. That question reminds me of the issue of the terminology in mathematics and its applications and of the cultural problems that arise in its use. We hope the mathematics is culture free, but applications, beginning with word problems, certainly are not.

I can't resist one anecdote here. In November 1968, Gail Young and I were sent to East Africa by the U.S. State Department. The State Department had been funding the Entebbe Project in mathematics education, and they wanted an evaluation. We found ourselves in Morogoro, an interior district capital about 100 miles from the coast of Tanzania, and we were talking to the management in the local girls' secondary school. The students had just taken the Cambridge examination, and we were told a tale of woe that has remained with me as an interesting example of cultural problems in applied mathematics.

Everyone there knew from past experience that the Cambridge examination usually had a problem on bearing. The teachers had drilled the girls thoroughly on every variety of problem on heading and bearing that could come up. Sure enough, the examination came, and the first problem was about heading. Not a student in the school tried it, even though all of them could have done it. Why didn't they try? Because the problem was written in terms of a traffic circle—called a "roundabout" in England and therefore in Tanzania. It even had a clover leaf! The students didn't know what the words meant, so they simply didn't try the problem. That exemplifies the cultural difficulty involved in making up an examination.

Of course, some of the cognitive science terminology that troubles teachers will also trouble mathematicians. If you say "metacognition," they are going to turn you off unless they understand what the word means. Of course, there's a professional terminology in mathematics, and we don't ever apologize for its not being understandable. Nevertheless, the terminology of the field of cognitive science must be made clear to the nonspecialist reader.

One of the problem-solving strategies that mathematicians often use, especially when they don't understand how to get from assumption to conclusion, is to derive whatever they can from the assumptions, no matter where it seems to lead. I am told that metacognition includes that kind of strategy. It seems strange to some mathematicians that "nonstrategy" is dignified by such an impressive word.

Steve Maurer's paper (chapter 7) reminded me of the importance of checking. Checking is a crucial strategy in mathematics at every level: Is what you are doing reasonable from every conceivable point of view? Success in mathematics has as a major component the instinct to think of more and more ways to check what you've done. Occasionally, in the old Bell Laboratories a research mathematician was hired and later fired. Why? He made mistakes, but that isn't why he was encouraged to leave. We all make mistakes. But he didn't have enough sense to check the special cases, the boundary conditions, every possible inconsistency. That's part of the strategy of doing mathematics that we must learn at every level. What can research tell us about how to learn that skill better? One of the Bell Laboratory mathematicians "solved" a well-known unsolved problem that he had been told was very difficult. He did it relatively easily in but a few pages. It didn't occur to him to ask himself why all those people had found it to be so difficult for so long. It happened that there was an algebraic mistake on page 2, which invalidated the whole development! How can we learn the judgement needed for deciding whether our results are believable?

Let me close with the question of problem formulation. The point of this example is that I don't want to formulate the problem, I want *you* to formulate it. The question will give you a feeling for the job of the person applying mathematics in the real world.

If you go into a supermarket, you will typically see a number of checkout counters, one of which is labeled "Express Lane," for x packages or fewer. If you make observations on x, you'll find it varies a good deal. In my home town, the A&P allows 6 items; the Shop-Rite, 8; and Kings, 10. I've seen numbers from 5 to 15 across the country. If the numbers vary that much, then we obviously don't understand what the correct number should be. How many packages should be allowed in an express line? I want you to take that question and make a well-defined mathematical problem out of it. First of all, students should discover how difficult problem formulation is. For example, we often say that we want to maximize this, minimize that, and optimize the other, and we forget that we can expect to do only one of those at a time. We have instincts about data that we are going to gather, but we have to decide what is relevant and what finally will be our optimization criterion. That's a typical problem out in the real world. I will act as the textbook at least to the extent of telling you that there is at least one beautiful formulation of the question I have posed. It is possible to devise a clear-cut mathematical formulation and to see what the data have to do with the problem and how you can actually solve it. Of course, I shouldn't have told you all that because, as I said earlier, your problem-formulating strategies may be changed as a result. But the question is still a good one.

REFERENCES

Reys, R., & Begsten, B. (1981). Teaching and assessing computational estimation skills. *Elementary School Journal, 82,* 116–127.

Reys, R., Rybolt, J., Begsten, B., & Wyatt, W. (1982). Processes used by good computational estimators. *Journal for Research in Mathematics Education, 13,* 183–201.

11

Cognitive Science and Mathematics Education: A Mathematics Educator's Perspective

F. Joe Crosswhite
Department of Mathematics
Northern Arizona University

I approach this topic as a consumer, not as a producer of research on cognition. I have been in mathematics education for over 30 years as a teacher, a teacher educator, and a sometime curriculum developer. During that time, few weeks have gone by that I have not been in a mathematics classroom. That has been my professional life. I think as a practitioner, and I value research to the extent that it informs practice. I intend my comments here to be constructive. If they seem to be mostly critical, it may be because I have not had time to frame them in polite language. Or, perhaps, my biases are so deeply rooted in the perceptions of a consumer that I am impatient with any research that seems at all remote from classroom practice.

I applaud the efforts of the organizers of this conference to bring together persons with different perspectives. They want to establish a dialogue with practitioners and have invited representatives from several disciplines and teachers from several levels to initiate that process. The discussions have reflected these different perspectives. But missing in the discussions so far and in the papers I have read is a sense of history—an historical perspective, if you will.

DEJA VU

When I first realized what this conference was about, I was reminded of my earliest involvement with the National Council of Teachers of Mathematics. My first committee assignment, in 1965, was with a research advisory committee that was being reborn after an absence of some 15 years. As a result of that assignment, I became involved in efforts to facilitate communication among researchers and between the research and teaching communities. One of the first outcomes of the NCTM committee was a National Conference on Needed Research in Mathematics Education held at the University of Georgia in September, 1967. That conference became known as the "Georgia Conference," and its proceedings (1967) were published as the first issue of the *Journal of Research and Development in Education* by the College of Education at that university.

The Georgia Conference was very similar in intent to this conference. It also brought together professionals concerned with research in mathematics education. Georgia was host to specialists in learning, teaching, and curriculum content. That conference also tried to stimulate dialogue among professionals with diverse perspectives, and to increase the probable impact of research on classroom practice. At that conference, and at the similar Greystone Conference 2 years earlier, research in mathematics education was often referred to as "an infant science." I was impressed then, as I have been here, with the frequency of statements like "We don't know enough." It is somewhat discouraging to hear again, after 20 years, that research in mathematics education may still be in its infancy. I wonder now if what we are seeing is not so much a case of perpetual infancy as the periodic rebirth of certain emphases in research—and if this new generation of research is simply the result of there being a new generation of researchers.

THE CONTINUING COMMUNICATION GAP

I encourage you to read the proceedings (1967) from the Georgia Conference. You will find discussions there that parallel the concerns addressed in this conference. In two fascinating papers (Davis, 1967a, 1967b) Robert B. Davis presents his view of the classroom teacher as an artist-practitioner.[1] He speaks specifically of a rhetoric gap between researchers and practitioners. He describes six types of rhetoric and notes that effective communication requires either that those forms of rhetoric be brought closer together or that a new rhetoric independent of them all be invented. I know the organizers and participants in the conference are sensitive to communication gaps. A way must be found to bridge those gaps if research is to have an impact in the schools.

---

[1]Some people view the teacher as scientist. There may be important differences in the expectations held for classroom teachers and, especially, in the way in which researchers try to communicate with them depending on the degree to which researchers subscribe to one or another of those points of view.

Channels for communication have improved over the past 20 years. The National Council of Teachers of Mathematics (NCTM) has sustained its commitment to research. The *Journal for Research in Mathematics Education* is in its 16th year; at least *it* has reached adolescence. Research sessions are now an accepted part of NCTM conferences. The special interest group (SIG) for research in mathematics education of the American Education Research Association (AERA) conducts a presession at each of the NCTM annual meetings. The Research Advisory Committee continues to focus Council attention on research. The *Arithmetic Teacher* and the *Mathematics Teacher* regularly invite articles interpreting research for classroom teachers. Mechanisms for communication do exist. Unfortunately, they are too seldom used to report research in a way that is palatable and instructive for the practitioner. The chasm between research and practice still exists.

Not only do cognitive scientists employ language that may be foreign to most practitioners, they seldom write for journals practitioners are likely to read. Neither, apparently, are they inclined to read or respect the journals and other publications for which practitioners write. At least that is my impression after reading a number of papers in preparation for this meeting. One long paper with a very promising title, "Implications of Cognitive Theory for Instruction in Problem Solving," had a bibliography of around 200 items—not one of which came from a publication classroom teachers would be likely to read. Yet there is a massive body of teacher-oriented literature in mathematics education, and it addresses issues of interest to cognitive scientists. Admittedly, much of it is not research based. But it does reflect insights born of experience. And it does contain the sort of practitioner's maxims that teachers are inclined to value. Moreover, it is usually written in the language of the classroom teacher, a language you need to understand. It is largely the omission of reference to this body of literature, both in the papers I have read and in the discussions, that makes me feel that historical perspectives are underrepresented in the deliberations of cognitive scientists.

In his paper for the Georgia Conference, Bob Davis (1967b) argued that "we live in an age when the best practice of the best practitioners almost certainly lies ahead of the best theory of the best theorists" (p. 59). That age may not have passed. It is important that cognitive scientists learn to talk to teachers. It is equally important that they learn to listen.

CYCLES IN MATHEMATICS EDUCATION

Just as concern with the communication gap tends to recur, other themes of mathematics education rise and fall in cycles. I sometimes refer to this tendency as a "cyclic pattern of overreaction" and describe it as an irregular sine curve with the horizontal axis a time dimension. Placing pairs of descriptors at the extremes of the vertical axis allows you to identify cycles. Although I agree with the writers of the NACOME (National Advisory Committee on Mathematics Education, 1975) Report that these cycles represent "false dichotomies" and unfortunate "polarizations of position," nevertheless, they do reflect shifts of emphases in school mathematics. For example, there has always been tension between process-oriented and product-oriented goals, between goals that serve the discipline of mathematics and goals that serve the social uses of mathematics. When disciplinary goals drive the curriculum, we move toward greater emphasis on rigor and precision in the mathematics we teach and on the processes inherent in mathematics and in mathematical thought. When social utility drives the curriculum, we have greater concern for basic skills and for activities motivated by our perception of the everyday uses of mathematics. In the 1970s, the back-to-basics movement was born of an overreaction to what was perceived to be an excessive emphasis on disciplinary goals in the new mathematics programs of the 1960s. Since then, school mathematics has been more product-oriented than process-oriented. We are still in or, at best, just emerging from that cycle.

The aspects of school mathematics most likely to be investigated by cognitive scientists relate primarily to process. Cognitive scientists need to be aware that many of the subjects discussed at the conference require more than modification of instructional strategies; they require modification of goals. And, historically, goals have not been modified on the basis of research. They change only as a result of shifts in philosophy. Instructional change becomes more difficult if the proposed goals are out-of-phase with prevailing emphases. Consider problem solving, for example. Problem-solving activities are certainly consistent with a curriculum focusing on social utility, but the instructional focus of a consumer-oriented curriculum may be more on the solution than on the process. The kind of problem solving discussed here would require a shift of emphasis for many teachers—cognitive scientists need to recognize that that shift may be contrary to current practice.

Problem formulation is one area that has been particularly sensitive to the cycles in mathematics education. I see little emphasis on problem formulation in school mathematics today. The kind of problem solving I see most is often "bottom line problem solving," taught through a method I describe as "bottom line teaching." It has been going on for at least as long as I have been in mathematics education. It went on even during the "new math" movement of the 1960s. Students learned to play a game with the teacher, a sort of gentleman's agreement. They sat quietly and seemed to be interested as the teacher explained "why," knowing that by the end of the lesson, the teacher would present a "how-to"—and that they would be held responsible only for that "bottom line." That is the kind of problem solving I see in most schools today. Students are waiting for the bottom line. They wait until the teacher reduces a set of problems to an algorithm and gives them a model by which the problems are to be solved. When researchers ask for problem formulation, and I think they should, they ask for a whole new problem solving environment. For most teachers that will represent a goal shift, a fundamental change in expectations. A change in instructional strategy is not sufficient, although it may be necessary. In this, and in other areas where existing goal structures are challenged, it must be recognized that that is what is being done. And strategies for intervention must be chosen accordingly.

ON METACOGNITION

I appreciated the discussion of metacognition more after it was reduced to "thinking about thinking." Then I could begin to understand what the discussants were talking about. I don't believe the word metacognition would have been used at the Georgia Conference. But it had an analogue then, and it has had analogues throughout the history of mathematics education. At one time we might have spoken of "rigorous thinking"; at other times of "critical thinking" or "logical reasoning." Even some connotations of "mental discipline" may not have been far removed from what is now being called metacognitive skills. In a booklet published over thirty years ago, and entitled *Thinking About Thinking* by the way, Cassius Keyser (1953) defined "autonomous thinking," considered it equivalent to "postulational thinking," and contrasted it with "organic," or "empirical," thinking. When you strip away the semantics, the kind of emphasis sought in mathematics education under the rubric of metacognition may not be all that different from what was sought earlier under its historical analogues. The relative amount of attention given to having students "think about their thinking" may just define another kind of cycle in school mathematics. There is certainly a substantial body of literature in mathematics education that addresses the rationale for and reports attempts to capture that sort of emphasis. Research on metacognition should refer to that literature, so that practitioners, especially those as old as I am, could better understand.

Research on learning and cognition surely has gone through cycles too. Most of us who are teaching mathematics, or teaching teachers of mathematics, were trained during an earlier cycle. If we relate at all to research on cognition, we relate to the language of that earlier cycle. I ask researchers to meet us where we are. Show an awareness of what has gone before. When you bring new terms into the lexicon, tell us how they are similar to or different from the terms with which we are familiar.

AN EXPERIMENT IN METACOGNITION

During the discussions of technology, the program GEOMETRIC SUPPOSERS was mentioned. The discussion of the higher order thinking skills that could be stimulated by that program also brought back the past. My doctoral advisor, Harold Fawcett, wrote a landmark document in mathematics education, *The Nature of Proof* (1938), which was published as the 13th yearbook of the National Council of Teachers of Mathematics. Fawcett intended to produce geometric supposers. That was the emphasis of his teaching. He did not talk about metacognition, but he thought a lot about thinking. His professional life focused on producing people who thought about their thought processes, who tried to understand their understanding. Surely that is close to metacognition.

Fawcett was greatly influenced by a statement made by E. H. Moore (1903) in his retirement address as president of the American Mathematical Society in 1902. Moore advocated that each student be directed to "set forth a body of geometric fundamental principles on which he would prroceed to erect his own geometric edifice." This would require that each student specify his or her own set of axioms, and it was projected that ensuing class discussions would make clear the functions of axioms in a theory of geometry and their relationship to theorems and definitions. Fawcett tried to put that into practice in the 1930s when he designed a course called "The Nature of Proof" in the laboratory school at the Ohio State University. I enrolled at Ohio State in 1961. Ultimately I became Fawcett's graduate assistant and had the responsibility of supervising student teachers working under teachers Fawcett had trained. I was discouraged by how little I saw of the philosophy he preached, even in geometry classes. Should that have disappointed me? Perhaps not. When Fawcett taught the course on proof, he used no textbook whatever. The students wrote their own texts, developed their own sets of basic assumptions, and invented their own theorems. Definitions evolved from a sort of class negotiation. Students decided how they would define a concept like "trapezoid" after discussing the consequences of alternative definitions. His classes must have been beautiful examples of the kind of thought processes we might hope to develop in mathematics education if we were emphasizing metacognition or one of its earlier analogues. But his approach could not be translated into classroom practice by most teachers.

Fawcett taught in a laboratory school. Even in the 1930s he may have had no more than three classes to teach each day. The proof course was taught over a 2-year period. Fawcett may have had a full-time graduate assistant. Teachers in Columbus, Ohio, in 1961 typically taught six or seven periods each day. They had no assistants. How could their classes have fully reflected the kind of thought Fawcett was trying to generate?

I remember well one student teacher who tried valiantly to capture Fawcett's philosophy. She taught geometry in a suburban school at a time when our student teachers had only a half-day teaching responsibility. She believed deeply in the Fawcett philosophy and had captured it very well—so well, in fact, that I had to make an emergency trip to the school for a conference with the student teacher, her cooperating teacher, and the principal. The cooperating teacher had complained to the principal that our student teacher was doing subversive things in mathematics, things like letting the students debate about whether they would accept the text definition. She even permitted—no, encouraged—students to go in different directions in their thinking about geometry! I had to try to explain why one of our students would do such unorthodox things.

In spite of that confrontation, the student teacher was hired in that same school the next fall. She was given a six-class assignment with four different preparations, was responsible for the school yearbook, and supervised the cheerleaders—all that in her first year of teaching. She no longer found time to do many of the creative, thoughtful things she had done as a student teacher. She could no longer individualize her instruction, allowing students to go in different directions in their thinking. She taught 1 year and quit. She told me that if she could not teach the way she was capable of teaching, she would not teach at all.

The problem in this situation was not with Harold Fawcett's philosophy, his theories, or his research. The problem was that we had not adequately translated his ideas into procedures that could reasonably be employed by a teacher working under the conditions that existed in the schools. We had not prepared our student teacher to capture the spirit of his research in ways that could reasonably be managed. My point here is that if research is to have an impact, even research as apparently close to the classroom as Fawcett's, a way must be found to put it into a format teachers can use. That may have to be a "ready-to-wear" format. Compromises may be required. Sometimes only selected elements of a desired model can reasonably be translated into classroom practice. Teachers, working under the loads they typically carry, cannot be expected to translate theory into practice on their own. A way must be found to bridge that gap for them.

PARTNERS IN RESEARCH

To maximize the probability that research will translate into practice, practitioners must be brought into full partnership. That partnership should begin even as you formulate the problems you intend to investigate.

I am reminded here of a prominent mathematics educator who directed a project on problem solving some 10 years ago. He was a very insightful, very competent mathematics educator. His project took him into elementary schools, where he conducted clinical interviews with students, talking with them about their problem solving. This was something he apparently had never done before. In his first meeting with the project steering committee, which included a number of experienced elementary school teachers, he exuberantly began to share the revelations he had uncovered in these long, intensive interviews. The typical reaction of the teachers was "Yes, that's the way it is." That almost destroyed him. He thought he had discovered the world, but those experienced teachers were already living in it.

Cognitive scientists must listen to teachers, ask them the research questions, share conjectures with them. Teachers may be able to save researchers a good deal of time and money by telling how it is—or how it isn't. They can surely help sharpen the formulation of research questions, especially if those questions are to relate to classroom practice.

Cognitive scientists may focus on a problem because it is both manageable enough to restrict the domain of their investigation and rich enough to supply the necessary examples. That focus is understandable. But they should not assume that the findings from this research will translate directly into practice. The problems that are of interest to cognitive scientists may not be the same as those that are of interest to classroom teachers. What may be psychologically significant may not be practically significant to teachers. The direct involvement of practitioners in research may help scientists to know when they are close to a problem of practical significance and help them design the translation studies that will bridge the gap between research and practice.

Teacher partners might also help researchers to know when they have something of value to report. Questions have been raised about the kind of research that is respectable among cognitive scientists and the credibility of research studies that use only a few subjects. The assumption seems to be that translation studies would necessarily precede taking clinical findings into the classroom. Generally, I suppose that should be the case. However, I suspect that cognitive scientists may tend to be too apologetic about what they "know." They need not always be so reluctant to share research results until they have dotted every "i" and crossed every "t." Teachers do not work at the 95% confidence level. If insights developed through research are rejected until there is virtual certainty of their validity, scientists may delay too long in sharing information to which teachers might respond most positively. Teachers are willing to take risks. If researchers have good ideas, they must find a way to share them even if the ideas have not been wrapped in what researchers would consider a final, perfected form.

In his discussion of the functions for cognitive technologies in mathematics education, Roy Pea described a research design he probably would not think of reporting now in an article for the *Arithmetic Teacher* or the *Mathematics Teacher*. It isn't a fully developed model; all the implications and ramifications have not been examined; it has not yet been subjected to a large-scale, controlled study to show its significant effects. And yet, I encourage Roy to put that model into a language that teachers can understand and share it with them. Now. There is value for teachers just in his conceptualization. If he can couple the model with some examples that are meaningful for teachers, they may find it useful even without subsequent confirming research.

Even bare intuition can be useful. During his presentation on problem formulation, Jeremy Kilpatrick held up a book written by Steven Brown. He identified it as the best source he knew on problem formulation. If we had been in a confessional, I would have asked Steve to tell us how much of his work was based on firmly established results from cognitive science and how much on intuitive insights developed through experience. If I know how things usually evolve in mathematics education, the intuition had more impact than cognitive science. Much of what we do is based on our intuitive insight. Sometimes that insight is informed by informal research, sometimes by research in progress. It may not always be the kind of research reportable in refereed journals, the kind of research that gets cognitive scientists promoted. But, to me, it is respectable research. It is based on careful observation of what is happening in classrooms, in negotiations between teachers and students. I have been impressed in the discussions here by how often many of you have observed that your research is confirming your intuitions. Capturing those intuitions in a format teachers can understand may be as important to mathematics education as scientifically tying them down in research terms.

Teacher partners may also help scientists understand when research has been carried beyond the level of detail they could reasonably expect to capture in a classroom. Research on cognition may generate as many new questions as answers. For each study completed, several new studies, each more subtle and more detailed, may be suggested. I am reminded of the research of one of my colleagues at Ohio State University on the role of positive and negative instances in concept formation. In his dissertation in the late 1960s, he found that the frequency of positive and negative instances did make a difference in teaching the concept under investigation. Since then he has extended his research to other concepts, has varied the sequence of positive and negative instances, and has tested their effects on different groups of subjects. He has uncovered a number of subtle differences in his research. He knows more about the role of positive and negative instances in forming concepts in mathematics than most teachers care to know, or, perhaps more accurately, than they could attend to in their classroom. Such levels of subtlety or detail have their place in cognitive science research, but researchers should recognize when the level of detail goes beyond what a teacher can use in making instructional decisions in the classroom. Having teachers as partners will help researchers to know when their findings may be translatable through teacher initiatives and when they must be embedded in instructional or curriculum materials.

STRATEGIES FOR CHANGE

We have spoken at some length about teacher education as a strategy for change. That is not only reasonable and desirable, it is necessary if research is to have any real impact in the schools. Only teachers can implement research findings. But are scientists patient enough to wait? And can they really expect the anticipated impact if they rely solely, or even primarily, on teacher education?

Preservice teacher education is an especially long-term strategy for change. First we have to educate the teacher educators! And then it would be another decade or so before we could reach even half the teachers in the schools. Inservice teacher education might secure more immediate results. But it has inherent difficulties too. Half the secondary school mathematics teachers in this country were in undergraduate programs before 1970. What sort of cognitive science do you suppose they studied? The turnover for elementary teachers is more rapid, but their numbers are out of sight. Even in the 1960s, when substantial funding was available for inservice programs for mathematics, we could reach only a small fraction of the teaching population. I believe that efforts should be concentrated at the inservice level and designed primarily for leadership personnel. And, I believe, research activity in teacher training should be in the specific context of instructional materials developed to incorporate research findings.

Data from the Second International Mathematics Study confirm earlier evidence that the predominant resource used by teachers is the basic textbook. They may not teach all that is in their text, but they do not go far beyond it. And how they approach a topic is greatly influenced by the approach in the text. They are also more likely to use supplementary instructional materials that are correlated with their text. So, in order to make an impact through instructional materials, researchers must either influence textbook authors and publishers to respond to their research, or write appropriate instructional materials themselves.

In my experience, research has had more of an impact on materials developed in funded projects than on commercial texts. The *School Mathematics Study Group* and the *Comprehensive School Mathematics Project* both purposely involved the research community and incorporated its findings into their work. There are no similar federally funded projects today, nor will there be, I think, in the near future. But there are several curriculum development projects underway that are supported by funds from the private sector. The major project at the University of Chicago seems very much inclined to respect research. The project developers intend to use psychological research, especially from the Soviet Union, and materials based on such research in developing the elementary school segment of their curriculum. I suspect they would be receptive to incorporating relevant cognitive science research.

Schools are searching for creative and productive ways to use technology, especially microcomputers, in mathematics instruction. If research in mathematics education is in its infancy, then some of the instructional software for mathematics that I have seen is embryonic. Developing good instructional software that incorporates theory and research findings and that is really usable by teachers might be a very effective and immediate way to affect practice in the schools. Because of adoption guidelines in several states, textbook publishers are under some pressure to provide software with their textbooks. They might be receptive to basing some of it on cognitive science. With the scientists' knowledge of and interest in technology, they may be able simultaneously to enhance the quality of available software and expedite the translation of cognitive science into educational practice.

OTHER WAYS TO MAKE AN IMPACT

During the discussion of bugs, one of the teachers commented, "I really don't care so much about where bugs come from; I want to know how to exterminate them." Cognitive researchers should care about how bugs got there. But they must go beyond that to reach practitioners. When they suggest that changing the curriculum sequence might forestall the development of bugs, they begin to talk my language. Then they have moved beyond identifying problems and are talking about solutions. Practitioners usually know the problems; they seek solutions.

Jane Martin pointed out that studies in the psychology of mathematics learning have concentrated on the early elementary level, where the problems, if not the subjects, are tractable. Cognitive scientists should not restrict their choice of research subjects that much, if they hope to maximize their impact on school mathematics. Students, curriculum, and instruction at the middle school level have too long been ignored in basic research. And yet, that level may be the locus of the most severe problems in school mathematics. Unless we make a serious effort to improve curriculum and instruction there, we will not begin to solve the problems of school mathematics at all. Researchers could make a major contribution by concentrating a good bit of their work in the middle grades.

I am encouraged by the algebra discussions. In that area, researchers are successfully getting at the interface between cognitive science and school mathematics. There is something new there and part of it is evolving from the technology. Cognitive science is investigating questions in a way that has not been possible in the past. The technology is not only giving us new information, it is also changing the questions. I hope work will continue in this area.

SUMMARY

In the foregoing pages I have suggested that historical perspectives may be underrepresented, if not in research, at least in efforts to communicate with practitioners. We ask researchers to meet us where we are, give us a starting point to know how their work is new and different, give us a frame of reference for understanding new terms they introduce or new constructs they propose.

I have also suggested that teacher education, alone, may be a slow and not fully effective process for change. Some research may be at a level of subtlety and detail that cannot readily be accommodated by teachers in the spontaneity of classroom interactions. Such subtleties need to be embedded in instructional materials. Ultimately and preferably, the aim should be to have an impact on basic textbook series. For now, concentration may have to be on developing well-designed and carefully related supplementary materials. The expanding demand for instructional software may offer an avenue for more immediate and possibly even more effective impact in the schools.

I have asked that researchers accept practitioners as full partners in their work. Not only will that expedite efforts to communicate with a practitioner audience, it may have salutary effects on the research itself. I have also asked that aspects of work in progress be shared. Researchers may have important things to say to teachers and other mathematics educators even before their research reaches a final, polished form. Intuitive insights are worth sharing too.

Although some research should be shared before that, much of what cognitive scientists do needs to be embedded in a sort of "translation research" that takes it closer to classroom practice. Practitioner partners can help researchers know when that is the case and help in the translation process. Sometimes compromises will have to be made. Researchers cannot expect that everything they like in a model can be transformed directly into practice. Sometimes they may even need to select research questions because they are essential to the translation process rather than because they are legitimate concerns of cognitive science. And, sometimes, scientists will need to test their clinical findings in the reality of the classroom, accepting the constraints that exist in that context.

Throughout these remarks, I have made far too sharp a distinction between researchers and practitioners. I should apologize for that. I did it to make a point but it is often an inaccurate and inappropriate distinction. It is particularly artificial for some of you I have known for years as mathematics educators, not as cognitive scientists. You are as much practitioners as I am. And others I have met for the first time have convinced me their interest in school mathematics is much more than a convenient context for their research. I think at heart they may be mathematics educators too.

REFERENCES

Davis, R. B. (1967a). Mathematics teaching with special reference to epistemological problems [Monograph]. *Journal of Research and Development in Education, Monograph, 1,* (1, fall).

Davis, R. B. (1967b). The range of rhetorics, scale, and other variables. *Journal of Research and Development in Education, 1,* 51-74.

Fawcett, H. P. (1938). *The nature of proof.* New York: Teachers College, Columbia University, Bureau of Publications.

Keyser, C. J. (1953). *Thinking about thinking.* New York: Scripta Mathematics, Yeshiva University.

Moore, E. H. (1903). On the foundations of mathematics. *Bulletin of the American Mathematical Society, 9,* 402-424.

National Advisory Committee on Mathematics Education (1975). *Overview and analysis of school mathematics grades K-12.* Washington, DC: Conference Board of the Mathematical Sciences.

Proceedings of national conference on needed research in mathematics education (1967). *Journal of Research and Development in Education, 1*(1).

Author Index

Subject Index

A

ABLE, 51–52
absolute value, 256
"accepting the given," 136
ACT, 137
AI. See *artificial intellligence.*
ALEX, 225
ALGEBRA ARCADE, 110
ALGEBRALAND, 81, 92, 111–114
algebra
 arithmetic basis of, 181–182, 255–256
 "bugs" in, 169–170, 176, 222–223
 equation solving. See also *equations.*
  Bundy's work on, 13–17, 226
  computer programs for, 103, 227–228
  "Generalized Substitution Principle,"
   232, 235, 238
  inversing procedure in, 228
  linear patterns in, 233, 235, 239
  mixed cues in, 230–231, 235–236
  planning in, 223, 234–235, 237
  quadratic patterns in, 234, 239
  radical as cue in, 233, 237
  rote vs. planned, 233–235
  Silver's work on, 226–227
  skill in, 231–232
  strategies for, 15–17, 229–232, 236–237,
   239, 243
  strategy errors in, 219, 223–224, 231, 236
  student difficulties with, 192, 218–219
 factoring, 150, 261

learning of, 181–182, 217–249
 analysis of preconditions, 226–227,
  237–238
 knowledge organization and, 220
 practice tasks and, 238–239, 241, 243
 problem sequence and, 221
 research in, 222–227, 246
 role of examples in, 225–227, 228–229
 student difficulties, 218–219, 223–224,
  247
 Wenger's theory of, 220–221
mathematicians' use of, 248
metacognition and, 192, 235, See also
 mathematical problem solving
problem schemata in, 47–48, 81, 83
word problems, 83
AM (Automated Mathematician), 132–133
arithmetic, 40, 110–111, 151–152, 181–182,
 247
ARITHMEKIT, 110
artificial intelligence, 10, 18, 130, 133, 179
 learning and, 51
 mathematics learning and, 166, 180–181,
  217, 222–228
  equation solving, 222–228
  planning in, 130
associationism, 2–3, 7

B

"back to basics," 6–7, 268
BANK STREET LAB, 105

285